PAPUA NEW GUINEA

Making a Living

Book 2

Pam Norman
Stephen Potek
Eron Hagunama
Joe Deruage

Oxford University Press is a department of the University of Oxford. It furthers the University's objective of excellence in research, scholarship, and education by publishing worldwide. Oxford is a registered trademark of Oxford University Press in the UK and in certain other countries.

Published in Australia by
Oxford University Press
Level 8, 737 Bourke Street, Docklands, Victoria 3008, Australia

First published 2005
Reprinted 2006, 2007, 2008, 2009, 2011, 2012, 2013 (third), 2014 (twice), 2015, 2016, 2017 (twice), 2018, 2020, 2023, 2025, 2026

ISBN 978 0 19 555877 7

Typeset by Midland Typesetters, Victoria, Australia
Printed in China by Golden Cup Printing China

Oxford University Press Australia & New Zealand is committed to sourcing paper responsibly.

Acknowledgment
This book was developed with the support of the Australian Government through the Curriculum Reform Implementation Project.

Contents

Grade 7

Grade 8

Secretary's Message

Making a Living is a new subject that aims to empower students with the essential life skills and knowledge to participate actively in society, to be self-reliant and to live sustainably now and in the future.

It combines the subjects of Commerce, Agriculture and Technology to promote useful and relevant life skills. It has been allocated 6 hours (360 minutes) per week to give teachers enough time to teach this very important subject.

Students will manage resources, promote better living and participate actively in community development. This subject depends on initiatives, dedication and commitment from teachers, students and the community. Its success will be evident through practical projects in the school and local community.

Making a Living is an important part of the reform curriculum that emphasises the need to teach our students to be enterprising and improve standards of living as called for in the Matane report: *A Philosophy of Education for Papua New Guinea.*

Students will investigate, create and design innovative ways of using available resources. They will make objects, cost them and evaluate the quality of their products and processes. This subject is relevant and useful for students leaving formal schooling after Grade 8 as well as those who will pursue studies in higher learning institutions.

Student Books 1 and 2, and the accompanying Teacher Resource Book, have been prepared to provide direct support for the achievement of syllabus outcomes. The content is consistent with, and promotes the teaching and learning models of, the syllabus and the related Department of Education Teachers Guide. These resources encompass a range of material including student activities, advice to teachers, assessment ideas, homework activities, project activities, reference information and a glossary of terms. They are intended to facilitate flexible learning activities for teachers and students, and may be modified and amended to suit local circumstances.

I commend and approve these materials for use in all Upper Primary schools throughout Papua New Guinea.

Peter M. Baki

Peter M. Baki CBE
Secretary for Education

1 GRADE 7
Managing Resources

Chapter Summary

In this chapter you will have an opportunity to:

- ✔ Investigate and compare the consequences of the mismanagement of land and water resources. In addition, plan a small project using appropriate management practices.
- ✔ Investigate and implement practical changes to reduce, reuse and recycle waste to benefit the local environment and community.
- ✔ Apply and demonstrate knowledge of crop management and animal husbandry practices by undertaking practical projects.

Syllabus References

Strand: Managing Resources

Substrand 1: Land and Water Management

Outcome: 7.1.1 Investigate and compare consequences of mismanagement of land and water resources and plan, design and undertake a small project using appropriate management practices.

Substrand 2: Environment

Outcome: 7.1.2 Investigate and undertake practical ways to reduce, reuse and recycle waste to benefit and improve the local environment.

Substrand 3: Crops and Animal Management

Outcome 7.1.3: Explain appropriate crop management and animal husbandry practices and demonstrate these through undertaking a practical project.

Land and Water

CONSEQUENCES OF MISMANAGEMENT

1. Loss of biodiversity and ecosystems SS

Communities of living things interact to support each other. These communities are called **ecosystems**. Sunshine, rain, soil, plants, animals and humans are all linked together in ways that enable living things to survive. Our ecosystem operates at five levels:

- Green plants convert energy from the sun into food in a process known as photosynthesis.
- Animals and insects that get their energy solely by eating green plants are called herbivores.
- Flesh-eating animals feed on herbivores. These animals are called carnivores.
- Some carnivores feed on other carnivores.
- Organisms, such as fungi and bacteria, break down dead or dying matter into nutrients so that it can be used again. These organisms are decomposers.

Biodiversity is the existence of a large number of different kinds of animals and plants that make a balanced environment. In Papua New Guinea we have a great diversity of insects, fish, spiders, mammals, birds, amphibians, corals, sea-life, plants and trees.

If humans don't manage land and water resources well, it can cause damage to ecosystems and a loss of biodiversity.

Birds of paradise would quickly become an endangered species if people were allowed to use guns to shoot them. The government's management strategy allows only nationals to hunt birds of paradise and this must be by traditional methods and for ceremonial purposes. This decision has been made so that the species survives. Similarly, crayfish with roe (eggs) should be left to breed and so replenish the reefs with more crayfish.

Both plant and animal species can become endangered or extinct. All human activity that impacts on the environment needs wise management. The loss of a tree, for example, has a significant impact on birds, animals, spiders and insects that depend on its leaves, branches, fruit, flowers, bark and roots.

A rainforest ecosystem—the plants, insects, birds and animals depend on each other to survive.

Some or all of these levels combine to form what is known as a **food web**, the ecosystem's mechanism for recycling energy and materials.

The survival of natural ecosystems is threatened by many human activities: clearing forests to make room for housing or agricultural land; damming rivers to harness energy for electricity; pollution of air, soil and water. Wise management encourages practices that enable humans to obtain necessary resources in ways that minimise damage to the ecosystem.

What food chains can you add to the following?

- flies – frogs – snakes
- plants – pigs – people
- plants – insects and small animals – birds
- algae and other water plants – small fish – sharks

For you to try

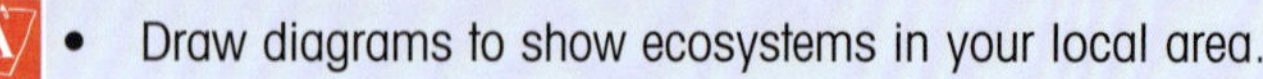

- Draw diagrams to show ecosystems in your local area.
- Discuss endangered species in Papua New Guinea.
- Go on an excursion to the forest in your local area and identify how the activities of people have affected land or water resources. Suggest how people could manage these resources better and write your suggestions in your journal.
- Plan a simple action that your school could do to protect an endangered species in your local area.

2. Soil erosion

Water, wind and human activity can erode soil. Loss of soil from water and wind erosion is balanced by the formation of new soil. However agricultural activity, logging, housing, industrial development, road construction and overgrazing by cattle destroys vegetation cover and greatly increases soil erosion.

Soil under and around buildings is important. Daily sweeping or raking of soil can lead to erosion. Soil erosion can undermine the foundations of buildings and concrete pathways. The result is damage to these structures and the need for repairs.

Good management is necessary to minimise soil erosion. Soil can be fertilised through crop rotation. The contour, or shape, of food gardens can minimise the amount of soil that is washed away. Covering a garden with mulch protects the soil from wind and water erosion.

For you to try

- Identify cases of soil erosion in your school or community. Write down what you think has caused the erosion and what has happened as a result.
- Write down all the things that cause soil erosion and suggest better ways to manage it.

3. Low food production

Mismanagement of land or water resources leads to low food production. Here are some mismanagement practices that lead to low food production:

- Allowing too many weeds to grow among food crops.
- Continuous burning to clear land leads to infertile soil.
- Using poor quality seeds or plant cuttings.
- Supplying insufficient water or too much water for plant needs.
- Destroying gardens during tribal fights.
- Failing to control pests such as locusts or mice.
- Over-fishing waters for commercial purposes.
- Failing to provide enough nutritious food to animals.

For you to try

- During a class discussion identify reasons why your school gardens could have low food production.
- Visit an area in your local community where you will be able to observe mismanagement practices that could lead to low food production.
- A Write a letter to the local council to suggest better practices, such as better fishing or gardening methods.

4. Water pollution SS

Thoughtful management is needed to maintain clean water supplies in rivers, coastal areas, lakes, bays, streams, underground water and household water tanks. Clean water is essential to support life on this planet. Mismanagement practices can cause water pollution and this can damage or kill living things. Polluted water is unpleasant to drink, look at, smell and swim in. Pollution can come from:

- Waste from the human body.
- Waste from animals.
- Fuel and oil spills.
- Dead animals.
- Household waste such as plastic bags, tins and tyres.
- Chemical waste from industry.

For you to try

- Go out into the community and find out what people do to have clean water for drinking and cooking.
- Discuss what people in your community do, that could lead to pollution of your school or community water supplies.
- A Investigate the extent of water pollution in your locality and identify the causes and consequences. Write about your findings in your journal.
- Plan a project to improve sanitation and waste-disposal systems in your area.

SS 5. Fewer resources

Poor management practices reduce the availability of resources.

Trees, for example, are a valuable resource in Papua New Guinea. They are used for firewood and for building houses and canoes. If cutting down trees is not well managed, there could be a shortage of timber for human needs.

Land is a valuable resource. If we use too much land for cash crops, there may not be enough land to grow food for local families. If too much land is used for industry, there may not be enough land for community needs. Land use needs to be managed wisely so that there is enough land to meet the needs of local people.

Minerals are non-renewable resources. The mining industry benefits the nation's economic development. However, as the minerals are exported, there are fewer resources available for future use. Mine closures greatly affect the lives of people in those areas.

Commercial fishing by local and overseas companies takes place in Papua New Guinea. Mismanagement of the fishing industry could lead to a shortage of marine resources for family or domestic needs.

For you to try

- Interview community leaders to find out which resources are less plentiful now than in the past.
- Go out into the community and observe and list sustainable management practices.

MISMANAGEMENT PRACTICES

1. Exploitation such as logging and fishing (SS)

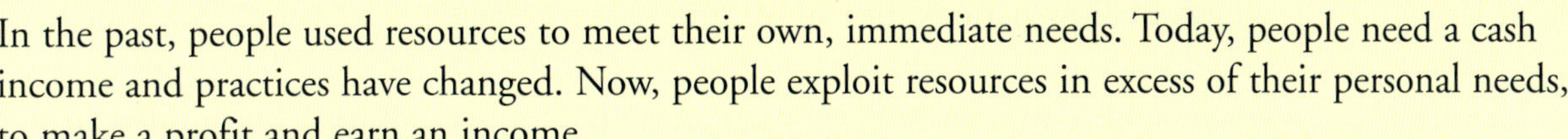

In the past, people used resources to meet their own, immediate needs. Today, people need a cash income and practices have changed. Now, people exploit resources in excess of their personal needs, to make a profit and earn an income.

The word **exploitation** means taking advantage of a resource to make a profit that is in the self-interests of a nation, company or person. Papua New Guinea wants investment from companies to develop logging, fishing and mineral resources. The exploitation of natural resources is necessary for economic industrial development. Local people want the income and advantages that come with industrial development.

However, mismanagement practices can bring short-term benefits and long-term losses. If activities are not carefully controlled, resources can become scarce or run out entirely. Companies require licences to exploit natural resources and activities are monitored by the government. Local people complain when they feel that companies are not operating within the law, and they complain if rewards for themselves and the nation are inadequate.

Forests have been cleared for large-scale plantation logging and agriculture. They have also been flooded for the construction of dams. Forests have been damaged by development and pollution related to mineral exploitation. Fishing companies exploit marine resources and some companies have set up canneries. Quality control of waters is important, to protect marine resources.

It is in everyone's interests to preserve the nation's valuable resources for future generations.

For you to try

- Invite a guest speaker to talk about the advantages and disadvantages of logging, fishing or mining activities in your local area.
- Identify and discuss exploitation of local resources to make a profit in your local area.

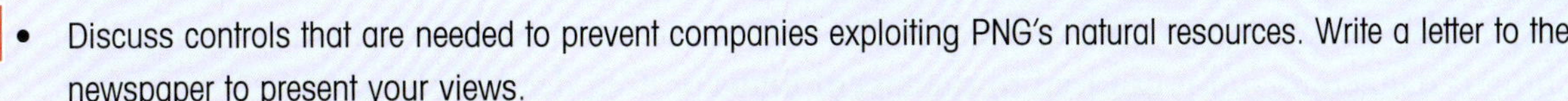

- (A) Discuss controls that are needed to prevent companies exploiting PNG's natural resources. Write a letter to the newspaper to present your views.

SS 2. Continuous cropping on the same piece of land

Continuous cropping on the same piece of land is not a good practice. Soil fertility is reduced after each harvest and the quality of crop yields worsens. Traditionally, farmers planted new crops on different land and this allowed fallow periods for soil to recover before being used again. However, as the population increases and some land is reserved for cashcrops (like coffee, tea, vanilla, copra, cocoa and oil palm), there is less land available for food cropping. It is now more common for families to practise continuous cropping on the same piece of land.

To overcome the problem of soil infertility, people need to learn and practise crop rotation and intercropping. For example, it is good to plant soil-depleting cereal crops, such as rice, and soil-restoring legume crops, such as peanuts and beans, on the same land, either one after another or at the same time.

Crop rotation is growing different crops, one after another, on the same land. This is better than planting onecrop only or a haphazard change of crops. The first crop could be corn, the second peanuts and the third taro. Crop rotation also conserves nutrients since the roots of the first crop may be near the surface and the second crop's roots may be deeper. In this way nutrients are drawn from different depths in the soil.

Another way to increase the organic content of the soil is to use **natural fertilisers** such as manure and compost. Mixing soil with animal wastes has been practised for many thousands of years and this creates organic compounds that are important for the growth of plants. Compost, which usually consists of dead plant and animal matter, works in a similar way to manure.

	Year			
Field	1	2	3	4
A	Corn	Peanut	Cocoa	Grass
B	Peanut	Cocoa	Grass	Corn
C	Cocoa	Grass	Corn	Peanut
D	Grass	Corn	Peanut	Cocoa

For you to try

- Plan a campaign to raise awareness about the disadvantages of continuous cropping on the same piece of land, and design a poster to promote crop rotation or the use of manure and compost.
- Visit gardens to observe continuous cropping on the same piece of land in your community and discuss ways to improve local practices.

3. Dynamite fishing

Dynamite fishing is illegal. It is an example of serious mismanagement of natural resources. Irresponsible people who want a quick and easy way to catch fish practise dynamite fishing. Dynamite fishing is dangerous and it can lead to the loss of limbs or even death of fishermen.

When dynamite fishing, explosives are placed in the middle of shoals of fish or reefs. The explosions are so strong that all the creatures within the explosion area are paralysed or killed. After the explosion, some fish float to the surface and others sink to the bottom. Fishermen collect the floating fish and sometimes they put on a mask and flippers to collect the fish that have sunk.

Dynamite fishing kills fish indiscriminately and is a serious threat to coral reefs. Fish habitat is destroyed and this prevents them from breeding. The end result is a devastated reef with far fewer fish for future harvesting.

For you to try

- Design a poster to illustrate the harm dynamite fishing can do.
- Suggest penalties for people who practise dynamite fishing.

4. Pollution such as inappropriate disposal of waste  

Mismanagement in disposing of waste causes pollution. A polluted environment is one contaminated by materials that reduce the quality of people's lives and their physical surroundings. Poor waste disposal threatens public health and it is unsightly. Waste threatens wildlife and contaminates rivers and ground water. Visitors are less likely to return to places that are littered with betel nut spit, plastic, old cans, discarded tyres and other industrial and household rubbish.

We need to improve our waste-disposal practices to ensure a clean, healthy and sustainable environment in Papua New Guinea. Waste products include:

- Agricultural waste: farm animal manure and crop residues.
- Ash: residues from the combustion of solid fuels.
- Dead animals.
- Industrial waste: chemicals, paints and sand.
- Large waste: demolition and construction debris and trees.
- Mining waste: slag heaps and refuse piles.
- Rubbish: non-decomposing wastes, either combustible (such as paper, wood and cloth) or non-combustible (such as plastic, metal, glass and ceramics).
- Sewage-treatment solids: material retained on sewage-treatment screens, settled solids, and biomass sludge.

Rubbish and waste encourages vermin, including rats, which carry plague, scabies and tropical diseases. Poor disposal of waste, especially containers, increases the risk of malarial infection. Still rainwater collects in old plastic bags or paint tins, and this is an ideal breeding ground for mosquitoes.

Disposing of waste is a widespread problem in Papua New Guinea. Waste is found scattered throughout towns and cities. In Port Moresby, village waterfronts are heavily choked with domestic rubbish adjacent to people's homes and on the roadway. Despite annual attempts to clean up, social attitudes appear unchanged and rubbish and waste quickly pollutes our environment.

For you to try

- Visit your local community to see positive, negative and interesting aspects of waste disposal in your community.
- Implement a plan of action to stop bad disposal of waste that pollutes your environment.

PD 5. Recycling, reducing and reusing waste

Wise management of 'waste' involves following the three Rs: **Recycle, Reduce** and **Reuse**. Instead of throwing things away, we should think of ways to recycle, reduce or reuse products. We can reduce the amount of rubbish we produce by becoming more aware of how much we throw out and by changing some of our habits.

To **recycle** is to reuse one resource to make another useful product. Old clothes, furniture, tyres, paper, cardboard, plastic, pipes, glass and aluminium are examples of materials that can be recycled.

- Children's clothes can be made from an old laplap or bed sheet.
- Wood from a damaged table could be turned into a stool.
- A plastic drink bottle can be turned into a plant holder.
- Scrap sheet metal can be made into a barbecue plate or cooking tray.
- Cardboard cartons can be made into storage boxes.
- Coconut shells can be turned into bowls, buttons, or necklaces.
- Rice and flour bags can be made into carry bags.
- Paint tins can be recycled as sawdust stoves.
- Small fuel drums can be turned into seats with a cushion on top.

To **reduce** involves making something smaller or using it less. Buy and use less of everything.

- When shopping, use bilums instead of plastic bags.
- Buy plain-packaged products to reduce flashy bulky packaging.
- Buy a large-sized product instead of two or more smaller products to reduce packaging.
- Start a compost heap to reduce the amount of food and garden scraps being thrown into your rubbish bin.
- Use cups and plates that can be washed and used again instead of paper or plastic items.
- Use cloth instead of disposable nappies for babies.

To **reuse** means to use a resource again, perhaps in a different way.

- Reuse empty glass jars as vases or storage containers.
- Reuse old car tyres as garden borders.
- Reuse plastic bags from stores to hold rubbish.
- Wash and reuse disposable plastic cups, plates or utensils, don't throw them away after one use.
- Repair things where possible instead of throwing them away to buy new ones.
- Hold a yard sale when you have things to get rid of; other people might find a use for items that you would otherwise throw away.
- Reuse gift paper when wrapping presents or covering books.

For you to try

- As a class talk about how the failure to reuse, reduce and recycle affects our environment. Is it worth the effort? Does it really harm our planet?
- Invite a health worker to speak about the dangers of waste to people and the environment.
- Conduct a class discussion to find out how you can reduce, reuse and recycle. Include individual as well as community activities.

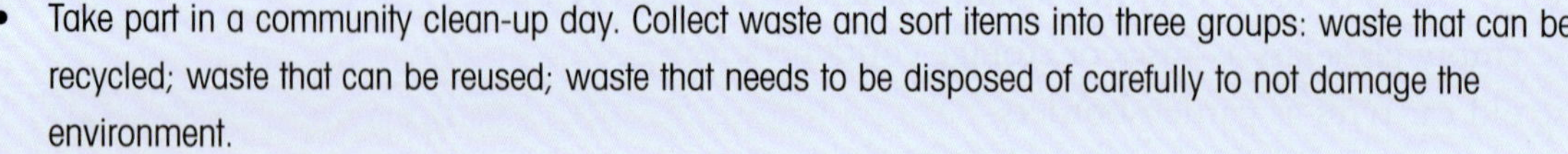

- A Take part in a community clean-up day. Collect waste and sort items into three groups: waste that can be recycled; waste that can be reused; waste that needs to be disposed of carefully to not damage the environment.

Class Project

Objective: Plan, design and undertake a small project using appropriate waste management practices. When starting out on any project the following steps should be taken.

1. ***Investigate***. Ask questions to find out current waste management practices in your local area and how they can be improved. This information can be collected by:
   - Writing a questionnaire and getting oral opinions from students and the local community.
   - Conducting a survey of the local community.
2. ***Plan and design***. Once the project has been determined, make an action plan. Organise students into groups and give each group a specific objective. At this stage students must be able to:
   - Write the procedures for completing the project as well as a time-line for completion.
3. ***Implement the plan***. Work cooperatively with the team to complete the task. Appoint a team leader who will:
   - Give every member of the team a task.
   - Make sure the project is progressing to plan.
   - Manage time and resources effectively.
   - Address problems if they should occur.
4. ***Evaluate.*** Once the project has been completed, it is important to evaluate the project.
   - How clear were the plans and procedures?
   - Did the students work as a team?
   - How reasonable was the schedule?
   - What were some of the problems that made it difficult to complete the project?
   - How will this plan improve waste management?
   - Can this project be extended or developed further?
   - Based on this project, what other suggestions do you have?
   - Write a report on the processes involved in the project.

Environment

WASTE AND RUBBISH: CAUSES AND EFFECTS 

1. Increased population and food consumption

The more people there are, the more food that is bought, caught, gathered, grown and eaten. Also, the more waste is created.

Here are some examples: tins from tinned fish or milk; plastic bottles from cordial and other drinks; glass jars from coffee; plastic shopping bags; cardboard boxes from matches or cereals; newspaper used to wrap store products; bones and vegetable peelings; and cloth, plastic and paper packets or bags from flour or rice.

Can we reduce, reuse and recycle these waste products to benefit and improve the local environment? Of course we can.

- Bones can be given to dogs to eat, thereby **reducing** waste.
- Glass jars can be **reused** as vases for flowers.
- Tins can be **recycled** and made into sawdust stoves or pots for plants.
- Large plastic bottles can be purchased instead of two or more smaller bottles, thereby **reducing** the amount of waste that is created.
- Plastic bags can be **reused** to hold rubbish or carry things.
- Vegetable peelings can be **recycled** as compost to enrich the soil.

For you to try

- Look at the contents of a home or school rubbish bin and list waste resulting from food consumption. Be creative in discussing ways to reduce, reuse and recycle the waste.
- Find out practical ways to reduce, reuse and recycle waste from food consumption to benefit and improve the local environment.

2. Hygiene and infectious diseases

Hygienic personal habits and a clean environment prevent the spread of infectious diseases.

Typhoid occurs when human faeces pollute food or water supplies. Early symptoms of typhoid are chills, high fever, headache, cough, vomiting and diarrhoea. If untreated, the disease develops into pneumonia, with intestinal bleeding, and it can be fatal. What can we do? We should ensure that we have hygienic toilet habits and clean hands when touching food.

Dysentery is a disease of the large intestine. Symptoms of dysentery are the frequent passing of watery faeces, often containing blood and mucus, and severe stomach cramps. Unhygienic conditions cause dysentery. The parasites and bacteria that cause dysentery breed in rubbish. They are spread by flies and ants. What can we do? We should keep our environment clean so that there is nowhere for parasites and bacteria to breed.

Malaria is transmitted by the bite of female mosquitoes. Symptoms of malaria include chills, shaking and fever. Mosquitoes breed in stagnant water. Water collects in old tyres, tins and other containers that are left lying about. What can we do? We should make sure that in our environment there are no rubbish products in which water can stagnate and mosquitoes breed.

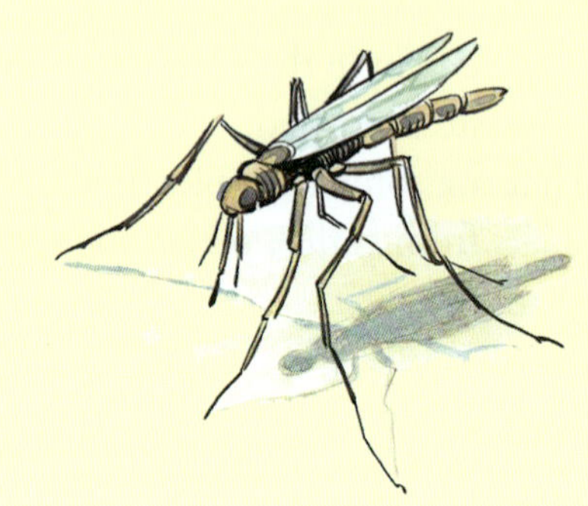

Hookworm larvae are able to penetrate the skin of any person who comes into contact with them. Infection is most commonly caused by walking barefooted in polluted areas or by handling human faeces. Symptoms of hookworm are anaemia, abdominal pain and diarrhoea. The disease also causes apathy (lifelessness) and malnutrition. What can we do? We must have hygienic toilet habits. We should not excrete on the ground where people may walk. Footwear should be worn in areas where people may have excreted.

For you to try

- Invite a health worker to speak to your grade about hygiene problems in your area.
- Plan an environmental hygiene program that looks at water supplies, sewage disposal and sanitary facilities.
- Find out and try practical projects to ensure that waste is disposed in appropriate places to prevent infectious diseases.

3. Consumerism

Consumerism involves the repeated purchase of an increasing variety of goods and services. Food, tools, axes, umbrellas, clothing, footwear, stationery, household items, electric goods, televisions, computers, sporting goods, entertainment equipment and vehicles are examples of goods we increasingly want to own. Advertising encourages us to update or improve the things we own. Perhaps families see the things they own as reflecting their status or success in life. This dependence on products and on constant consumption defines our modern consumer society.

Obviously, the more things we purchase, the more we create waste and rubbish products. The more things we purchase, the more opportunities we have to reduce, reuse and recycle. When we shop, careful planning enables us to purchase wisely and consider packaging and waste. We can reduce the number of plastic bags we use. We can buy in bulk to reduce unnecessary packaging. If we can reuse a jar or tin, for example, we should choose this style of packaging over a cardboard packet. We should also write a shopping list to reduce impulse buying.

For you to try

- Visit the local market. Identify waste that results from purchases at local markets. Describe how waste from market purchases could be reduced, reused or recycled.
- Draw pictures of items that a family might buy each week. For each item write sentences to identify the potential waste and describe how it could be reduced, reused or recycled.

- Develop a list of guidelines to encourage wise shopping and write this list in your journal.

4. Excessive packaging

Packaging is the wrapping or material around a product. The purpose of packaging is to contain, protect, identify and facilitate the sale and distribution of a product. Nearly all manufactured and processed goods require packaging. The basic materials of packaging include paper, cardboard, cellophane, steel, aluminium, glass, wood, cloth and plastic. Excessive packaging is when too much or unnecessary material is used.

- **A shirt** may be packaged in a cardboard box with a cellophane see-through front. Many pins are used to hold the shirt in place around a sheet of cardboard inside the shirt.
  Is all this packaging really necessary? Would you buy the same shirt if it were on a hanger in a shop?
- **Toys** are often presented in colourful attractive boxes. A small baby's rattle or a model toy car, for example, may be packaged in a comparatively large box. The packaging is costly and adds nothing to the usefulness or quality of the product.
  If you had a choice, would you buy the same product in or out of a box?
  We should assume that the product would be cheaper if it was not in the box.

Packaging contributes to environmental pollution. Millions of tons of packaging are discarded as rubbish each year. Some packaging is biodegradable but most is not degradable. Non-degradable material does not decompose or decomposes very slowly in the natural environment.

In villages we may burn or bury our rubbish. In towns, the council has a collection service to take rubbish to landfill areas. Here bulldozers compact the rubbish into layers and cover each layer with clean soil, which is also compacted. Another way of dealing with the problem is to recycle packaging materials as new products, packages or fuel.

For you to try

- Visit a shop. Identify items that you feel have too much packaging. Give reasons why you think this is so. Suggest different ways of packaging. Explain your ideas to the class.
- Design three good ways of packaging a product for sale.

BENEFITS TO THE ENVIRONMENT

1. Clean air and water

Clean air and water are necessary to life on earth. Good waste management practices ensure clean air and water in our environment. Papua New Guinea is a developing nation and care must be taken to control air and water pollution to keep us and the environment healthy.

Harmful substances in the air damage the environment, human health and reduce our quality of life. Dust, exhaust fumes from vehicles, gases from factories and rubbish dumps, and smoke and ash from fires pollute the air. Air pollution makes people sick. It causes breathing problems and lung damage. It harms plants and animals and the ecosystems in which they live. In some countries people wear masks over their noses and mouths to filter the air they breathe.

Substances that pollute water supplies are harmful to living things. People who drink polluted water can become ill. Seafood from polluted waters may be unsafe to eat. Plants and animals cannot survive if their water is loaded with poisonous chemicals or harmful germs.

Many human activities produce substances that pollute water. Fertilisers and pesticides from agriculture find their way into water systems. Polluted water can flow from factories and mines. If untreated, human sewage and garbage pollute water systems. Oceans become polluted from careless action. Plastics thrown overboard can kill birds or marine animals by entangling them, choking them or blocking their digestive tracts when swallowed.

For you to try

- Invite a speaker from the health department to talk about air, water and waste concerns for the local area.
- Collect water from three different locations in your community and check its cleanliness. Which water could you drink and which could you not drink? Give reasons for your answers.

2. Healthy environment

To maintain a healthy environment, we must clearly understand that the Earth's resources are limited. We need to reduce the amount of resources we use and reduce the amount of waste we create. We need to reuse and recycle products to conserve the resources we have. Avoid developing a 'throw-away' mentality. The future growth of Papua New Guinea depends on living in ways that protect the environment while meeting the basic needs of our people.

A healthy environment means everything around us must be clean and disease free. To make this happen, everybody must be responsible for his or her actions.

- Care for rubbish bins so that dogs do not scatter the rubbish nor vermin breed.
- Burn waste that can be burnt.
- Dig rubbish holes to bury waste that will not burn.
- Use organic waste in compost heaps.
- Keep fresh water supplies free from contamination.
- Be thoughtful and cooperate with council waste collection services.

Develop personal hygienic habits:

- Be thoughtful in how you use rivers or bush land for toileting.
- Do not spit betel-nut juice so that it spoils the environment.
- Dispose of rubbish thoughtfully.
- Do not throw rubbish out of boats or vehicles when travelling.
- If you are tempted to drop rubbish on the ground, think, 'Who will pick it up?'

It should not be someone else's responsibility to pick up your rubbish. Take the time to put rubbish in an appropriate place. The health of your environment is your responsibility.

For you to try

- Look around your classroom, school and school grounds to assess the standard of cleanliness. Write rules for waste management to care for your school environment. Make a poster displaying these rules.
- Select an area and pick up rubbish. Find out the source of the waste materials and develop a plan of action to reduce the problem.

3. Reduction in disease PD

Good waste management benefits the environment by reducing the chances of disease.

- Mosquitoes cannot breed in tyres or tins if they are buried in the soil. This reduces the chance of getting malaria.
- Parasites and harmful bacteria cannot breed in a hygienic environment. Hygienic conditions reduce the chance of getting dysentery.
- Human faeces will not contaminate food and water supplies if people have clean hands and hygienic toilet habits. The chance of getting typhoid is greatly reduced if human body wastes are disposed of in a hygienic and appropriate manner. Modern toilets should be used where these are available.
- Wear shoes and take care when walking on ground that people may have used as a toilet. Hookworm larvae breed in such areas and readily penetrate people's feet. The larvae move through the body and hook onto the stomach walls where they take the nutrients from food. People then suffer malnutrition, anaemia, stomach pain and diarrhoea. The chance of getting hookworm is reduced if toilets, and not the bush, are used to dispose of human body waste.

For you to try

- Arrange for a health worker to talk to you about the diseases in your locality that are caused by unhygienic conditions. Plan questions to ask.

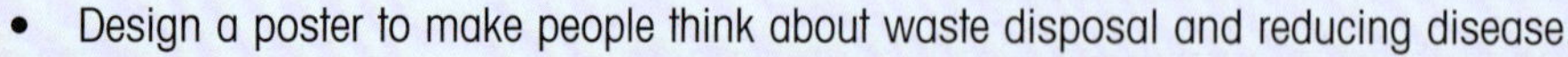

- Design a poster to make people think about waste disposal and reducing disease.

4. Reduction in pollution

Our environment benefits from all efforts to reduce pollution, both big and small. Careless disposal of waste products in our towns, and along roads and beaches, is a serious problem. Betel nut spit is both unsightly and unhygienic and it is a common sight throughout Papua New Guinea. Plastic bags and cellophane wrappers are thoughtlessly discarded and spoil the appearance and health of our environment. Papers, tins and bottles are dropped. Market waste-matter piles up and overflows from bins. Old vehicles rust in yards, settlements and villages.

- **By reducing the waste created, we reduce pollution.** Next time you go shopping, think carefully about packaging. By choosing something with less packaging you are helping to reduce pollution.
- **By reusing products, we reduce pollution.** Before you throw something out, think carefully about whether or not you can find another use for it. Perhaps you could use an old exercise book to press flowers. An old saucepan or bucket with holes in it could become a plant holder.
- **By recycling products, we reduce pollution.** You could make old newspapers into fuel blocks. Old rims from bicycle wheels can be turned into toys for children. A stand for a rubbish bin can be made from recycled timber or metal.

For you to try

- Plan an income-generating project by recycling something that would otherwise be waste in your locality.
  1. Talk about waste that could be a resource for recycling.
  2. Develop an action plan.
  3. Make the recycled product.
  4. Market the product and evaluate the outcome.

TYPES OF WASTE ITEMS

1. Old furniture

Look around your home, school and community to find old furniture that is not being used. You may find a table, chairs or cupboards made of local or imported materials such as wood, metal or plastic. Think of ways to recycle or reuse the old furniture. You need to be creative. Look at the construction materials and imagine other possibilities.

We could reuse the furniture by repairing it and so maintain its original purpose. By sanding the surfaces, stitching fabric and repainting or varnishing items, we can recreate attractive furniture to be sold or used again. This is better than discarding the old furniture as waste.

We could recycle the old furniture in other ways. Wood that is damaged could be used as firewood or as garden stakes. Better quality wood could be made into stools, storage shelves, boxes, bed frames, roadside stalls or bookshelves. Metal frames or legs can be cut and recycled to form other products such as barbecue racks, gates, plant stakes, fishing spears or handles for spades and rakes.

For you to try

- Find out where to find old furniture in your local area. Get permission to repair or recycle the furniture.
- Discuss the different ways you could use the old furniture.

- Develop and implement plans to repair and sell these items to generate funds for the school.

2. Waste products generated from locally produced and imported goods

Different types of waste are produced from imported and locally produced goods. These waste products include plant waste, coconut husks, sawdust, off-cuts from timber yards, tins, bottles, tyres, paper, cloth, rubber and plastic.

Waste product is generated when working with logs. At timber mills, logs are cut into planks and beams for construction work. In the process, sawdust and off-cuts are generated as waste products. In a similar way, as people make canoes and kundu, or garamut drums from logs, wood chips and off-cuts are generated as waste product.

We can reduce waste from logging by reusing or recycling sawdust, wood chips and off-cuts.

- Sawdust and wood chips can be used as ground cover in chicken yards. After the chickens are sold, chicken manure mixed with sawdust can be dug into garden soil to enrich it.
- Sawdust can also be used in sawdust stoves as a cheap source of fuel.
- Off-cut pieces of timber can be used to construct tabletops, seats, market benches, *haus wins*, chopping boards or carved artefacts. Items can be made for personal use or for sale.

Flour and rice bags can also be reused or recycled.

- Bags can be reused for carrying garden produce to markets, carrying garden waste to a dumping area or storing items at home.
- Bags can be recycled and made into floor mats, clothing, purses or book bags.

We are limited only by our imaginations. We can find many new uses for different types of waste products generated from imported and locally produced goods. As few jobs are available in the workforce, we need to be clever in devising ways of making a living from resources available in our community. These resources must include products that we would otherwise regard as waste.

For you to try

- Go around your local community and observe and list different types of waste from imported and locally produced goods such as tin cans, plastic bottles and old tyres.
- Discuss how this waste can be reused and recycled for personal use or for sale.

3. Origins of familiar products

What are the origins of roofing iron, bath towels, newspapers, tin cans, car tyres, plastic bags and glass bottles?

- **Roofing iron** is made from sheets of iron or steel that are galvanised (coated with a thin layer of zinc) to protect the base metal from wearing away. The sheets of iron or steel are first dipped in acid to remove dust, dirt and grease. They are then washed and dipped into molten zinc. It is common to see old sheets of roofing iron being reused or recycled.

- **Bath towels** are made from cotton fibres that are spun into threads and woven into towels. Small trees produce cotton. The flower bud naturally splits open to reveal a mass of white fibres 1.3 to 6 cm in length. Cotton is light, strong and absorbent. This makes it suitable for towels, nappies, bedding, bandages and tropical clothing. It is common to see pieces of fabric, rags, old towels, nappies and clothing being reused or recycled.
- **Newspapers** are made from paper. Most paper is made from wood pulp. However, about five per cent of all paper is made from cotton rags and linters (the short threads on cotton seeds). The wood chips or cotton rags are mixed with a solvent and pounded until the fibres separate and form a smooth pulp. This is spread on a belt of fine mesh that passes between a series of heated rollers that dry and compact the pulp into sheets of paper. It is common to see paper from newspapers and magazines, as well as wrapping paper, reused or recycled.

- **Tin cans** are made of aluminium or very thin steel. Because it is light in weight, resistant to rust, easily shaped and suitable for use with food and drink, aluminium is widely used for cans. Recycling aluminium containers is important for saving resources and energy.
- **Steel** is made from iron ore. Steel cans, such as paint tins or fuel drums, are stronger and more rigid than aluminium cans. It is common to see tins and cans of all sizes reused or recycled.

- **Car tyres** are made from natural and synthetic rubber mixed together with oils, waxes and fabric. The gummy mixture is layered and moulded into tyres. It is common to see reused or recycled car tyres.
- **Plastic bags** are made from light, strong material that is produced by chemical processes and can be formed into shapes when heated. There are many different types of plastic. Plastic bags are made from materials that come mostly from oil, coal or natural gas. Plastics are fairly inexpensive to produce, lightweight, waterproof and easy to colour. They do not rot or rust and can be clear or opaque. Plastic bags are made by blowing hot air into a plastic tube and inflating it, like a balloon, until a bag of the desired shape and size is formed. It is common to see plastic bags littering the sea and ground and spoiling the environment. It is very important that we look for ways to reduce, reuse or recycle plastic bags.

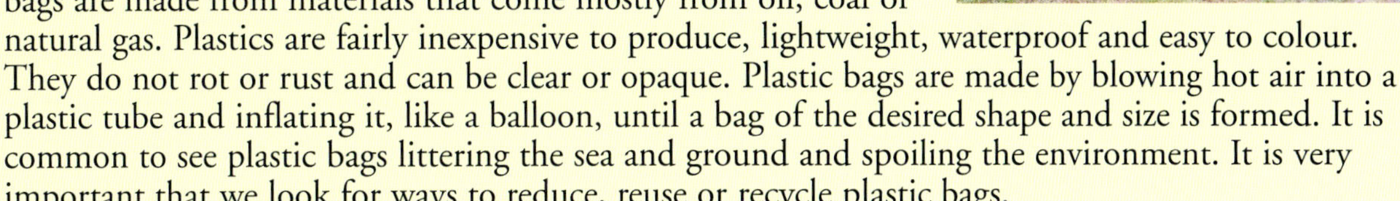

- **Glass bottles** are made from glass. The basic ingredient of glass is silica that comes from sand. This is mixed with other materials and heated to a molten state. Casting, blowing, pressing, drawing and rolling are ways of shaping molten glass. Glass has many uses; bottles, windows, drinking glasses, vases and light bulbs. It is common to see glass products being reused or recycled.

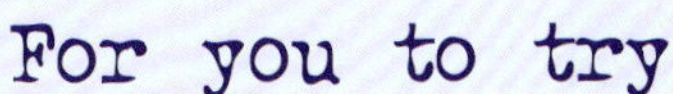

For you to try

- Bring a familiar product from home to school. Write a short description of its origins. Display the product and your description. Give an oral presentation to other students explaining how the product could be reused or recycled.
- Look for information in library books, or from other sources, about the origins of other familiar products.

4. Production processes that create waste

Different types of waste are created when making or producing things. This waste may be sludge, liquid, gas or solid. Some waste can be reused or recycled; other waste cannot. The important thing is to be thoughtful in protecting the environment.

Hazardous waste comes from mining operations. As chemicals are used in the mining process, run-off from tailing ponds can pollute rivers. This damages fish habitat and can also pollute water supplies. Slag heaps of waste product build up on land. Mining companies should have plans to minimise the environmental damage caused by waste from mines.

Making copra creates waste products. The leftover coconut shells and husks can be burnt as fuel. Coconut husks can also be used as scourers. Fibres from coconut husks are sometimes used to stuff mattresses. Coconut shells can be made into bowls, buttons, earrings or hair combs.

Water waste needs to be treated before it can flow harmlessly into the earth or a river. In a treatment plant, the waste water is passed through a series of screens, chambers and chemical processes to reduce its harmful effects. Many homes have septic tanks where waste water flows from a buried outlet into rock-filled trenches below the earth's surface. The waste water soaks into the soil.

Chemical waste can come from home activities. Chemical wastes include paints, aerosol cans, paintbrush cleaners, glues, turpentine, varnish, batteries, battery acid, motor oils, brake fluid, petrol, diesel, kerosene, fluorescent tubes, photographic chemicals, insect sprays, nail polish, nail polish remover, shoe polish, medicines, fertiliser and rat poison. All of these are hazardous and they should be disposed of thoughtfully to protect the environment.

For you to try

- Identify products that are made in your locality. Draw a flow chart of the processes involved.
- Identify types of waste that could be a by-product of the process. Discuss what should happen to the waste products in the best interests of the environment.

Class Project

- Investigate and undertake practical ways to reduce, reuse and recycle waste that is generated from imported and locally produced goods such as old tyres, water tanks, large containers and newspapers.
- Plan an income-generating project by recycling something that would otherwise be waste in your locality.
- Investigate and undertake practical ways to reduce, reuse and recycle waste from food consumption to benefit and improve the local environment.

Crops and Animal Management

APPROPRIATE CROP MANAGEMENT PRACTICES

1. Land preparation SS

It is important that land is well prepared before planting. Healthy soil is necessary for good crop yield. Plants get water, oxygen and essential nutrients from the soil. Not all soils have enough nutrients. For this reason we should add green weeds, manure, compost and other fertilisers to the soil. Organic matter improves drainage and helps to aerate the soil. Air and water can then more easily pass into the soil. Both air and water are needed for healthy growth of roots and soil organisms. Water in soil contains dissolved plant nutrients that the roots absorb.

Before planting crops, land should be prepared as follows:

- Cut down weeds with bush knives and put them in a compost heap.
- Dig out big stumps, dead logs, shrubs and wild ferns.
- Turn over the topsoil with hoes and garden forks.
- Spread manure or compost evenly over the soil.
- Bury the manure or compost in the soil using hoes or spades.
- Green weeds can also be dug into the soil to add humus.
- Smooth the surface soil with a rake or spade.

For you to try

- In your locality, examine land that is used for cash crops and food crops. Find out about the land preparation methods that were used. Write a short report on your findings.
- Identify land at or near your school that can be used for cash or food crops. Prepare the land for planting using appropriate skills and tools.

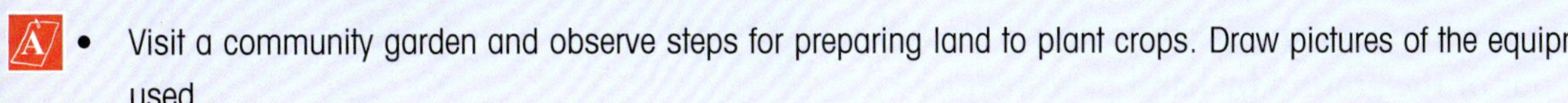

- Visit a community garden and observe steps for preparing land to plant crops. Draw pictures of the equipment used.

2. Planting methods: nursery and direct planting

Two planting methods are to:

- plant seeds in a box-tray in a nursery, or
- plant seeds or cuttings directly into the soil.

A **nursery** is a shelter under which seed trays are placed on benches. A nursery allows seeds to germinate and become young seedlings in an area that is protected from heavy rain, wind, insects and weeds. Tiny seeds for cabbages, lettuces, carrots and tomatoes can be started in this way. Seeds are sown on or close to the soil surface. They develop into tiny plants or seedlings and they can then be transplanted into the prepared garden soil.

Crops such as corn, pumpkin, beans, peas and peanuts have bigger seeds and these can be planted directly into the prepared garden soil. They are sown 3-4 cm below the soil surface. A small hole is made and the seed dropped in. It is gently covered with soil to keep the seed moist. It can then germinate and develop its root system.

Crops such as taro, yam, sweet potato, sugar cane, cassava, aibika and pitpit can be grown from cuttings that are planted directly into the soil. A small hole is made; one end of the cutting is placed in the hole and gently surrounded by soil. Roots form on the part of the plant that is covered in soil. Stems and leaves form on the section of the cutting that sits above the soil. In this way a new plant is generated.

For you to try

- Visit gardens in your local area. Observe and discuss planting practices for both food and cash crops. Observe planting methods.
- Visit plant nurseries in your area. Set up a nursery at your school that will generate income for the school.
- Collect suitable seeds and cuttings to plant in the prepared garden areas of your school. Compare germinating plants in nurseries with direct planting.

3. Weeding

Weeds are plants that grow where they are not wanted. Weeds are not wanted in cultivated areas because they compete with crops for sunlight, water and soil nutrients. Also, many weeds carry organisms that infect crops and cause disease.

One way of removing weeds is simply to pull them out of the ground by hand or to remove them using a spade or hoe. It is important to remove the underground roots of weeds, together with the tops. Weeding should be done when the soil is moist as the roots are more likely to stay attached to the upper part of the weed.

Another way to control weeds is by mulching. Covering soil with mulch prevents weeds from germinating. Mulching also adds nutrients to the soil.

Another way to control weeds is to use chemical weed killers. However, weed killers can pollute the water and soil, and kill crop plants, so they need to be used with great care.

For you to try

- Invite a gardener to your school to explain why weeding a crop garden is important. Identify problem weeds that grow among crops in your local area.

- Go into your school garden and identify weeds that can be used in compost or dug into the soil and weeds that carry disease that you should burn.
- Try weeding in both moist and dry soil. Compare the effort involved and the results.

4. Mulching

Mulching is a good crop management practice. Mulch may be straw, sawdust, shredded tree bark, leaves, coconut waste, grass or manure. It is spread in a thick layer over the soil. Mulching prevents weeds from growing and water from evaporating from the soil. It keeps the soil cool for seeds to grow. As soon as the seeds have germinated, the mulch must be moved away from the stems to give the young plants light and air.

As well as protecting plants, mulch decays in time and enriches the soil. This is important when crops are grown one after another. Digging in the decayed mulch feeds the soil and this creates a better environment for growing new crops.

For you to try

- Go to your local market and collect plant waste that can be used for compost.
- Prepare mulch mixtures and spread them in thick layers over soil and around plants. Explain the reasons for mulching.
- Invite a local agricultural officer to talk about mulching practices of families in your community. Evaluate practices and suggest improvements if necessary.

Coconut mulch

5. Irrigation (SS)

Irrigation means to supply water to land by artificial means. In Papua New Guinea, crops are most often irrigated using buckets or water hoses. This takes a lot of time and human energy. All countries of the world irrigate when rain does not fall regularly enough, or in sufficient quantity, to grow crops. Irrigation ensures plant growth and crop yields. Four methods of irrigation are flood, channel, sprinkler and drip.

Flood irrigation is suitable for growing rice where the ground is level and water is plentiful. Water is allowed to flood the ground for a given time, depending on the crop, the type of soil and its drainage. Paddy rice, grown in flooded fields, is common in Asian countries.

Channel irrigation involves making water channels between crops that are grown in rows. There could be a dam or river beside a field and when necessary, a barrier is removed to allow water to flow along the channels between the rows of plants. When enough water has reached the crops, the barrier is replaced.

Sprinkler irrigation places sprinklers at regular intervals along a pipe. Water droplets spray out of each sprinkler in a circle. When water has reached the roots of the plants, the sprinkler system can be turned off. Sprinkler irrigation uses less water and has finer control than channel or flood irrigation methods.

Drip irrigation: long lengths of narrow plastic tubing, with holes at regular intervals, are laid over the ground where tree crops are planted. Usually attached to a timing device, small but frequent amounts of water trickle on to the ground close to the roots of each plant. This method also uses less water and has better control than channel or flood irrigation methods.

For you to try

- Invite a local water board officer to talk about irrigation systems in your local area.
- Generate ideas for an irrigation system that could be developed for cash or food crops grown at your school. Seek advice from an agricultural officer. If possible, install your irrigation system.
- Compare the benefits of an irrigation system with watering crops by hand.

6. Harvesting

Harvesting is the time when crops are gathered. If you harvest good quality and quantity crops, you have used appropriate crop management practices. To growers, quantity and quality are equally important. They both determine how much a crop is worth. You need to harvest crops when the time is right, not too early and not too late. Farmers also try to harvest crops at a time when prices are high. Traditionally, special festivals are held at harvest time.

Crops such as tomato, pawpaw and bananas are harvested when the fruit is ripe. Crops like aibika, lettuce, cabbage and spinach are harvested when the leaves are mature and tender.

Growing rice

Corn and beans are harvested when they mature. Root crops, like carrots, are harvested when the root is a good size but before it becomes woody and loses its sweetness. Potatoes, yam, cassava and taro are harvested when the tubers are needed and they are a good size.

Cash crops such as coffee, rubber, cocoa, coconut and vanilla are also harvested in particular ways. Rubber is harvested by collecting latex from v-cuts in the trunks of the rubber trees. Vanilla pods, coffee beans and cocoa pods are picked from trees when they are ready. Dry mature coconuts for making copra are gathered from the ground after they fall from trees.

For you to try

- List a variety of cash and food crops grown in your locality. How do people know when the crop is ready to be harvested?
- Invite a local person to explain how a traditional calendar is used by the community to harvest crops.
- Visit a plantation at harvest time and listen to a grower talk about ways of harvesting crops.
- If you have cash or food crops at your school, describe how you will know that the time is right for harvesting.

7. Processing

Food processing describes all the stages that food goes through from the time it is harvested to the time it is sold.

Processing may simply involve picking, sorting and washing fruit and vegetables before they are taken to market.

Other processing methods convert raw materials into a different form or change the nature of the product. For example, sugar is made from sugar cane, jam from fruit, flour from cereal or root vegetables and sago is made from sago palms.

Cash crops such as coffee, rubber, cocoa, coconut and vanilla require complex processing before they are sold.

For you to try

- Write to a company to get pamphlets to find out how cash crops are processed.
- Discuss and write a report on the processing of a food crop in your school garden, from the time it is harvested until it is ready to be sold.

8. Storage

Most often food crops are not stored; they are eaten or taken to market as soon as they are ready. This means that when produce is plentiful, prices drop. When produce is scarce, prices go up. One way to make fruit and vegetables last longer is to store them at low temperatures.

Some crops can be successfully stored for a long time before they are used. Leaf bundles and clay pots are used to store sago. Pandanus and cane baskets are used to store potatoes and onions. Hessian sacks are useful for storing sweet potatoes and pumpkins. In the Trobriand Islands, yam huts are built to store yams. Crops should be dry before they are stored. Mould and rot will quickly spoil food.

Insects, rats, mice and mould are enemies of stored produce. All food crops should be stored in clean dry places and checked regularly for damage. Damaged food should be removed to stop the disease or rot from spreading to other food.

Class project

Design and prepare a storage area for crops from your garden or for food you use in your cooking lessons. Describe your storage area.

9. Pest and disease control

Fungi and bacteria attack plants and cause disease. When you first see the signs of diseased leaves, look for ways to get rid of them very quickly.

In Papua New Guinea, common pests are insects, beetles, locusts, snails, weevils, rats and mice. Insecticides and poisons can be used to control pests but these are poisonous and can affect humans. For this reason, food should be washed before being cooked or eaten.

- As you walk around the garden, pick off and destroy any insects that you see.
- Destroy all rubbish and places where insects could hide, feed and breed.
- Change crops so that the same plants are not always grown in the same place.
- Digging your garden sometimes exposes larvae to the sun and they die.
- Plant your crops at the right time of the year when there are fewer insects.
- If you feed and water plants, they will be more resistant to insects.

For you to try

- Invite an agricultural officer to talk about pests that can ruin crops in your local area.
- Go into community gardens and look for signs of disease or damage. Find out the cause of the problem and seek advice on how it should be treated.
- List ways to control pests and diseases in crops grown at your school.

SS APPROPRIATE ANIMAL HUSBANDRY PRACTICES

Animal husbandry is the agricultural practice of breeding and raising livestock.

1. Housing

Many types of animals are kept for personal use or to generate an income. These include pigs, ducks, chickens, goats, rabbits, cassowaries, sheep, cattle and horses. While animals can be free-range (allowed to wander freely), housing provides protection from weather and harm.

Housing for livestock should:

- Provide shelter from rain, wind, heat and cold.
- Be well ventilated.
- Let in enough light.
- Be easy to clean.
- Provide protection from thieves.
- Provide protection from other animals.
- Provide enough room for each animal.
- Provide access to food and water.

Animal housing can be varied in shape and size. It can be made from different materials. The type of housing provided will depend on the livestock being kept, resources available, and the locality and climate. Bush materials, such as thatch roofs, bush timber walls and packed earthen floors, may be used in some areas. Iron roofs, milled timber or chain wire walls and concrete floors may be used in other areas. Walls for pig houses need to be stronger than walls for chickens. In the highlands, closed-in walls protect animals from cold. Coastal animal housing can be more open.

A deep layer of straw or sawdust covers the floor of some animal enclosures. This collects animal waste and can be used later as compost for gardens. Special feeders and water containers can be purchased for chicken houses. A flow of air is necessary to minimise bad smells and help keep the area dry.

For you to try

- Invite a livestock farmer to explain the types of housing he constructs for his animals.
- A Visit a livestock farmer to observe animal housing. Write a report, with labelled diagrams, on the animal housing you observed.
- Demonstrate your knowledge and skills by designing and building a house for a chosen animal, either at your school or for a family in your community.

2. Feeding

To be healthy, farm animals need food and water. Water should be available at all times. If left to roam freely, animals forage for the water and food they need. If you watch chickens, ducks, pigs or goats feeding around villages, you will see them eating leaves on plants, digging for worms, scratching for seeds and seeking other food.

Animals kept in captivity cannot find their own food, so the food they are given must provide a balanced diet. The amount of feed depends on the age and type of animal, the quality of the feed and the weather.

Some animals, like goats, sheep and cattle, graze on plants. Special grasses are sometimes planted to provide rich pasture for grazing animals. In captivity they are fed stalks and grain from crops such as corn, barley, oats and sorghum.

In addition to food from foraging, pigs can be fed vegetables from the garden, cooked rice and food scraps from the house.

Chickens and ducks can forage freely or be fed food scraps. Also, commercially produced mash can be purchased. It is a prepared mix of grain crops such as corn, barley, oats and sorghum. 'Chicken Starter', 'Growers' Mash' and 'Layers' Mash' are different types of feed for chickens.

For you to try

- Visit a chicken farm to find out about different types of chicken feed and their cost.
- Investigate the design of feeders that prevent feed from being spoiled or wasted.
- If resources are available, set up a small chicken project in your school to generate an income.

3. Choosing breeding stock

It is essential to choose breeding stock with care. Domestic pigs and chickens look very different from their wild ancestors. Popular animal breeds are the result of selective breeding practices.

Imagine that you are looking at six pairs of pigs, each pair is a different breed, and you are required to choose one pair for breeding purposes. How would you make your choice? Imagine that you are looking at six pairs of chickens, each pair is a different breed, and you must choose one pair for breeding. How would you make your choice?

Choice of breeding stock may be determined by: the availability of breeds, quantity, quality, cost, personal preference, climatic conditions, veterinary services or feeding requirements. Desired characteristics include: appearance, age, sex, egg-laying ability, quality of meat produced, rate of growth, resistance to disease, absence of physical defects and fertility. By choosing to breed from stock

with desirable characteristics, we promote the survival of selected breeds and the possible extinction of other breeds.

Scientists are concerned that selective breeding reduces biodiversity and threatens the survival of some breeds. Conservation programs for threatened domestic breeds of cattle, sheep, horses and pigs have been established in some countries.

For you to try

- Visit a livestock farmer who has breeding animals to find out how he selects breeding stock.

4. Health and hygiene

Appropriate animal husbandry practices protect the health and hygiene of animals and prevent animal diseases. Farmers are concerned about disease for several reasons:

- Disease can reduce the productivity of animals that are raised for slaughter.
- Sick animals can affect the economic well-being of many industries.
- Some animal diseases can be passed on to humans and control of these types of diseases is vital to public health.

Mad Cow Disease and Asian Bird Flu are two animal diseases that caused a major international health crisis. In a number of Asian countries, millions of chickens were slaughtered in response to an outbreak of Avian Influenza—commonly known as Bird Flu. This was very bad for Asian farmers. In Britain, thousands of cattle were slaughtered as a result of Mad Cow Disease. This seriously hurt cattle producers and the beef industry. Strict quarantine regulations aim to prevent animal diseases from other countries spreading to Papua New Guinea.

Animal diseases may be infectious or non-infectious. Parasites, bacteria and viruses cause infectious disease when they enter an animal's body. Non-infectious diseases can be inherited or caused by diet, environment or injury. Veterinarians are doctors who treat animal health problems.

People often say that prevention is better than cure. But what can a farmer do to prevent diseases infecting his animals? Any farmer can take the following action:

- Provide quality housing that protects animals from harsh weather conditions. It must have good ventilation.
- Make sure animal housing is clean and disinfected before bringing in new stock.
- Keep food and water supplies clean. It is important to regularly clean and maintain containers.
- Manage floor litter to prevent the build-up of disease organisms.
- Provide plenty of good quality feed as malnourished animals more easily get sick.
- Store feed in proper containers.
- Manage pest problems such as rats and mice.
- Always buy disease-free animals.
- Separate sick animals from healthy animals to prevent sickness spreading.
- Seek advice from a veterinarian (a doctor who treats animals) if health problems persist.

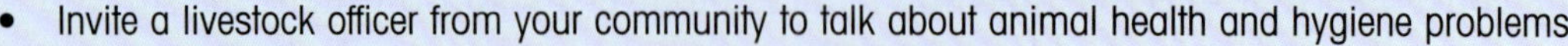

For you to try

- Invite a livestock officer from your community to talk about animal health and hygiene problems.

- Prepare and display posters that give advice on appropriate animal husbandry practices.
- If you have animals at your school, draw up action plans to promote the health and hygiene of your animals.

5. Handling

We need to take care in handling animals to prevent injury to animals or ourselves. Common injuries include being bitten, scratched, butted, knocked down, kicked or pinned between an animal and a hard surface. We should never relax when handling animals as they behave in unpredictable ways. Animals react unexpectedly when you remove them from their home.

- Wear protective clothing when handling animals. Long trousers and footwear help guard against injury.
- Wear rubber gloves when handling sick or injured animals.
- Wash your hands thoroughly after handling animals.
- Keep animal areas clean and free of rubbish and sharp objects.
- Remove anything that could make you (or the animal) trip or slip.

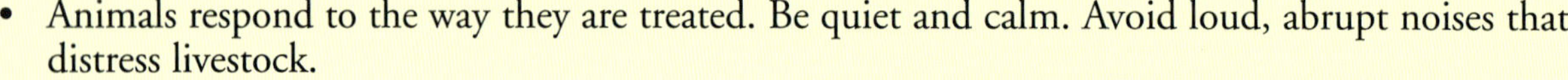

- Animals respond to the way they are treated. Be quiet and calm. Avoid loud, abrupt noises that distress livestock.
- Pens should be equipped with a gate.
- Animal areas should be evenly lit as animals can panic when moved from a light area to a darker area.
- Moving or flapping objects can cause problems and make animals hesitant to move.
- All livestock tend to refuse to walk over any change in ground texture or surface, for example a drain, grate, hose, puddle or shadow.
- Livestock with young are usually more defensive and difficult to handle. When possible, let the young stay as close to the adult as possible.

All these factors need to be considered when planning livestock handling facilities.

For you to try

- Visit a livestock farmer to observe how workers handle animals, for example, loading pigs onto trucks or putting chickens in portable cages to take them to market. Discuss the methods used, the level of distress caused to animals and the safety measures employed by the handlers.
- Discuss and list appropriate handling practices for livestock at your school or in your community.
- Identify problems or dangers in handling animals and ways to avoid them.

6. Slaughter and processing

To 'slaughter' means to kill an animal and 'processing' refers to the steps required to prepare animal meat. All slaughtering should be done quickly to avoid undue suffering by the animal.

Killing chickens or ducks can be done quickly by breaking the neck. Hold the bird by its legs. Place your other hand behind the head. Pull down, twisting to the right. This breaks the neck. The bird will flap its wings for a few seconds, but it will soon be still. Straight after killing, hang the bird by its feet and pluck out the feathers. Plucking is easier when the body is still warm. If the bird is left for a while, dip it in hot water to make the plucking easier. After plucking, cut the head off. Cut around the vent taking care not to cut into the rectum. Remove the intestines, liver, heart and lungs. Rinse the bird out with cold water. After these processes, the bird is ready for cooking. The bird can be stored in a freezer if you do not want to cook it straightaway.

Stunning is a technique used to kill pigs in Papua New Guinea. Animals are bled after stunning and they die from loss of blood before they regain consciousness. Use a sharp knife to slit the pig's throat on clean ground. Slit the throat while the heart is still beating so that all the blood gets pumped out. Make sure you have severed the main veins and arteries. Wash the pig down.

For you to try

- Arrange a visit to a nearby village to observe the slaughtering of a pig, goat or chicken.
- Draw up a table of instructions for slaughtering and processing animals that will help you in later life.

Better Living

Chapter Summary

In this chapter you will have an opportunity to:

- ✔ Discuss and analyse aspects of a nutritious diet and identify different ways of obtaining, processing and preserving food.
- ✔ Identify and assess required maintenance of home and school buildings and take appropriate action.
- ✔ Investigate consumer rights and responsibilities, and develop skills to become a wise consumer.
- ✔ Initiate plans and apply techniques and processes to design items that will benefit the individual as well as the local community.

Syllabus References

Strand: Better Living

Substrand 1: Healthy Living

Outcome: 7.2.1 Analyse aspects of a nutritious diet and suggest how and where students might obtain, preserve, process and prepare foods to meet nutritional requirements.

Substrand 2: Care and Management

Outcome: 7.2.2 Assess home and school buildings to identify areas that require maintenance, repair or other improvements and take appropriate action.

Substrand 3: Wise Consumer

Outcome: 7.2.3 Investigate consumer rights and responsibilities and demonstrate practical ways to be a wise consumer.

Substrand 4: Making Things

Outcome7.2.4 Initiate plans and apply appropriate techniques and processes to design and make an item that should benefit an individual or the community.

Healthy Living

ASPECTS OF A NUTRITIOUS DIET

1. Balanced meals that use food from the five food groups 

Foods contain different substances called **nutrients**. The main nutrients are **protein, carbohydrate, fats, vitamins, minerals** and **water**. Some foods are rich in one nutrient but have little or none of another. For example:

- Cooking bananas have large amounts of carbohydrate but very little protein.
- Fish and meat have good supplies of protein and iron but no carbohydrate.
- Fruits have lots of water and are a good source of vitamins and sugar but they have very little protein.

It is important to eat a variety of foods each day to get a 'balance' of nutrients.

Growth	Energy		Protective	
Protein	**High energy**	**Staples**	**Fruits**	**Vegetables**
Fish: fresh/tinned Meat: fresh/tinned Nuts Winged beans Eggs	Coconut milk Oil Margarine Pork fat	Taro Cassava Sago Sweet potato Rice Cooking banana Flour	Avocado Guava Bananas: ripe Pawpaw Pineapple Mango	Dark green leaves Cabbage Carrots Pumpkin Corn Tomato Capsicum

For you to try

- Draw a table and list foods from each of the five food groups that are available in your area.
- Plan and prepare 'lunch packs' that are balanced meals, for example: leaf parcels of fish, rice and vegetables; egg sandwiches with a piece of fruit.
- Using the above information, devise a class project to sell nutritious school lunch packs at affordable prices.

2. Preparation methods that maintain foods' nutritional value

Staple foods like rice and potato need to be cooked to soften the starch cells. Meat also needs to be cooked to soften the fibres. Peanuts are better when cooked, either roasted or boiled. When cooked, these foods can be digested more easily and the nutrients absorbed.

Fruit and coloured vegetables lose nutritional value from heat, drying and storage. Fresh is best. If coloured vegetables need to be cooked, use as little water as possible and cook for a short time. Stir-frying vegetables is a quick way of cooking coloured vegetables. It keeps them crunchy and maintains their nutritional value. Some vitamins and minerals are water-soluble. Always try to use vegetable water in soups and stews.

Fruit is more nutritious if it is eaten fresh and not cooked. The vitamin C in tomatoes, guavas, pawpaw, pineapple and oranges is destroyed by heat and storage.

Clean hands and equipment, as well as packaging, prevent good food being spoilt by germs Remember: to maintain the nutritional value of food, cleanliness is essential.

For you to try

- **Make coleslaw.** Wash and finely shred cabbage, carrots and spring onion. Add some peanuts and moisten with coconut cream.
- **Make stir-fried vegetables.** Wash and finely shred cabbage, carrots, beans and spring onions. Add some peanuts. Lightly fry and sprinkle with soy sauce.
- Suggest two ways to prepare each of these foods: potatoes, eggs and pawpaw. Choose three more foods to add to the list and suggest food preparation methods. Choose one food and list the advantages and disadvantages of each preparation method.

PD 3. Nutritional requirements based on age, sex and level of activity

The nutrients each person requires differs according to:

- Age (young children or adults)
- Sex (male or female)
- State of health (well, sick or recovering from sickness)
- Activity level (hard work or light work).

Cocoa pods

Fruit or cooked food for young babies needs to be mashed, as they have no teeth. Food for old or sick people may need to be soft if they find it difficult to cut or chew.

Children are active and growing so they need balanced meals. They also have small stomachs so children need smaller meals than adults. Because children go to school, they need balanced meals to be packed for school lunches. Lunches can be contained in leaves, coconut shells, bamboo, clean paper or plastic.

Women who are pregnant or breastfeeding babies need calcium, iron and protein-rich foods. Calcium is needed for strong bones, iron for healthy blood and protein builds the body. All are essential for mother and baby to be healthy. Eggs, milk and meat all contain these nutrients.

People who are active and do hard physical work need energy-rich food. A house builder needs more energy food than an office worker. For morning tea, a scone or biscuit and a cup of tea, with sugar, would give added energy to the builder.

Fruit is good for people with colds or infections. Oranges, lemons and guavas are rich in vitamin C. Vitamin C helps the body resist infection and it heals body tissue.

People who are pale and get tired easily may be anaemic. They need meat and dark-green leaves which are rich in iron.

Children with marasmic-kwashiorkor need extra carbohydrate and protein foods. Children with marasmic-kwashiorkor have a big belly and thin muscles in their arms and legs.

For you to try

In groups, plan meals for the following groups of people:

- Six-month-old babies
- Primary school-aged children
- Sick, elderly grandparents.
- Pregnant women
- House builders

HOW AND WHERE TO OBTAIN FOOD

1. Commercially produced food

Many of the foods that we eat are made on a large scale in factories, bakeries and food-processing plants. These include sugar, rice, flour, tinned meat, tinned fish, margarine, cooking oil, bread, hard biscuits, sweet biscuits, milk, coffee, tea and cordial. We buy them in small and large shops.

Bread from a bakery

The ingredients are weighed and poured into a high-speed mixer and blended into soft dough.

The dough is cut into loaves.

The loaves are put on trays until they increase in size.

The bread is baked in a hot oven and then cooled.

Some bread is sliced by machine. The bread is wrapped by machine, ready to be sold.

For you to try

- Draw a flow diagram to show the process for making two food products that you buy from shops.
- At each stage of production, what checks should be made to ensure that the product is safe to eat and of good quality?

2. Exchange or barter system

In the past, people would get food by exchange or barter systems. For example, a clay pot could be exchanged for a bunch of bananas; a pig from one clan could be traded for fruit and vegetables from another clan. The people involved would need to agree that the items were of equal value. In modern society, the exchange of money has replaced the barter system.

For you to try

- Find out if the elders in your community used exchange or barter systems to obtain food.
- Ask questions to gather information about items that were traded for food.
- Discuss advantages and disadvantages of a barter system.

3. Home-grown products

Home-grown products are widely available throughout Papua New Guinea both for personal use at home and for sale in markets. These are fresh, readily available and provide value for money. Home-grown food products include:

- **Protein foods:** chicken, seafood, bush animals, peanuts, pandanus nuts, winged beans, and eggs.
- **Fats:** coconut oil, pork fat.
- **Carbohydrates:** yam, sago, taro, sweet potato, sugar cane, and coconut.
- **Fruit:** bananas, pawpaw, pineapples, guava, mangoes, custard apples, laulaus, lemons, oranges, passionfruit, and avocado.
- **Vegetables:** a wide variety of green leafy vegetables, pumpkin, carrots, broccoli, choko, tomato, corn, and peas.
- **Water:** young coconuts.

It is becoming common to see cooked foods for sale in markets. Scones, doughnuts, sago, banana cakes and smoked fish are examples of cooked foods available.

For you to try

- Draw a diagram of the area around your family home. Show any food that you produce. This could include animals (like pigs or chickens), food bearing trees (like mangoes, lemons or coconuts) as well as your vegetable garden. Label your diagram. Display and compare diagrams.
- Discuss the advantages of home-grown products.
- Identify which home-grown products are available at your school for **Making a Living** lessons. Discuss what more could be done.

4. Sharing

Another way of getting food is through sharing with friends and relatives. Sharing is common when one person or family has more food than they require for their own needs.

Perhaps there are too many ripe mangoes for a family to eat. Maybe many fish were caught on a fishing trip. Perhaps there is too much meat on a pig for one family to eat. Maybe there is enough sago from a sago palm for more than one family. These are some reasons why there could be extra food available.

If there is no way to preserve the extra food or save it from going bad, it is best to share the food with others while it is still good. Sharing food with others is better than letting good food go to waste. Through sharing food we help others to have a nutritious diet.

For you to try

- When did you last share food with someone else?
- What did you share? Why did you share it?
- What are the advantages and disadvantages of sharing?
- Discuss occasions when others have shared food with you.

PRESERVING FOOD

Preserving food keeps it in a useful condition and prevents it from going bad. Food can be preserved in many ways – drying, pickling, chilling and salting. There are many reasons for preserving food:

- Extra food can be preserved for later use.
- Preserved foods add variety to a family's diet.
- Home-made preserved food is cheaper than bought preserved food.
- Preserved foods can be sold to generate income.
- Crops that do not grow all year round can be preserved, stored and eaten in times of shortage.

1. Smoking food

You can preserve fish and meat by smoking them. Cut the fish or meat into pieces. Place on a rack about 60 cm above a fire. Direct the smoke at the fish or meat. The warmth of the smoke cooks the food. Smoked food will last much longer than fresh food. The smoke also gives a distinctive, pleasant flavour to the food.

For you to try

- Smoke some fish or meat.
- Calculate a selling price for a piece of smoked fish or meat.
- What could you combine with the smoked food to make a tasty and nutritious meal or school lunch?
- Consider smoking food to generate income.

2. Drying food

Fruit and vegetables can be sun-dried. When drying food be sure that the food is protected from moisture and insects. Sweet potato, beans, corn, peanuts and pumpkin can be dried for use at a later date. Dried starchy food can be ground into flour.

Dried, thinly sliced fruit, such as banana, mango and pawpaw, make delicious snacks. The drying process concentrates the sugar content and the fruit slices become crisp sweet snacks. They can be sold in small, plastic-wrapped packages.

Sun-dried fruit

Peel and thinly slice fruit.
Spread on trays and dry in the sun.
Cover with netting if insects are a problem.
Turn the fruit once a day.
When dry, store in plastic bags or a screw-top container.
Consider selling packets of dried fruit-chips.

3. Pickling food

Pickling is a way to preserve vegetables, such as cucumber, onion or green pawpaw. The vegetables are preserved in vinegar and later used to add flavour to food. The acid in the vinegar prevents the growth of microorganisms.

Fresh green pawpaw pickle

Peel and remove seeds from one medium-sized pawpaw.
Shred pawpaw on a coarse grater.
For each cup of pawpaw add ½ cup vinegar, 1 tablespoon of sugar and spices to taste. (Spices may be 1 small chilli or 1 teaspoon of chopped ginger)
Bottle and store until ready to serve.
Use as a pickle with cold meat or hard-boiled eggs.

For you to try

- Make fresh pawpaw pickle. Use with slices of tinned meat on savoury biscuits or sandwiches.
- Ask children in other classes to taste the pickle and decide whether it could be sold.

4. Chilling food

Chilling and freezing are methods of preserving food at temperatures that are low enough to prevent decay. Foods that are frozen in a freezer will keep for a long time. Food kept in the chilling part of a refrigerator keeps for a shorter time. When you visit a large store, notice the types of food kept chilled and the types of food kept frozen.

For you to try

- **Class activity.** Draw the outline of a refrigerator on a large sheet of paper. Show where you would store ice cream, bottles of water, margarine and coloured vegetables. Draw or cut food pictures from newspapers and glue them on to the chart to show where they should be stored. Explain why.

5. Salting food

You can preserve fish by salting it. Salt kills microorganisms and stops their growth. You will need a large container like a plastic bucket and one kilogram of salt for each kilogram of fish. Clean and fillet the fish. Rub salt on all sides of the fish. Place a thick layer of salt at the bottom of the container. Make layers of fish and salt until all the fish is used up. Fish can be left in salt for a long time without spoiling. Before cooking salted fish, soak the fish in water overnight to remove excess saltiness.

PROCESSING FOOD

1. Making juice

Refreshing and nutritious drinks can be made from the juice of fruit.

- To make orange or lemon juice, squeeze the juice from the fruit and add water and sugar to taste.
- To make banana or pawpaw juice, mash the flesh of ripe fruit and add water and sugar to taste.

For you to try

- Experiment using different types of fruit to make juice drinks.
- Substitute coconut juice for water.
- Make ice blocks by freezing fruit juice.

2. Canning food

Almost any kind of food can be preserved in sealed cans. In shops you will see canned fruit, vegetables, milk, meat and fish. Canned food gives variety to the family diet. Canned meat and fish are popular in Papua New Guinea. They provide a convenient source of protein when fresh food is not available and when people do not have refrigerators. Canned fruit and vegetables are more expensive and less nutritious than fresh garden produce.

Canning takes place in factories. Papua New Guinea has meat and fish canning factories. Always check that the can is in good condition. Check the use-by date on cans. Never buy cans with rust or blown ends as this indicates that the food inside could be spoiled.

For you to try

- Collect empty cans of food purchased by your family. Wash them and make a display in your classroom. Read labels on cans and talk about the contents and nutritional value.
- Compare the cost and nutritional value of preparing food from the garden with the same product in a can.

- Draw a flow diagram to show the process of making a canned product.

COOKING FOOD

Most food needs to be cooked to make it soft and easier to digest. Cooking is the process of preparing food by heating it.

Moist heat cooking methods are boiling and stewing. Water, or some other liquid, is added to the food in the cooking process.

Dry heat methods are frying, baking and grilling. Oil may be used in dry heat cooking methods. Different methods of cooking are used to add interest and variety to meals.

For you to try

- Bring some taro, sweet potato or other type of potato to school. Boil, bake and fry.
- Compare the tastes of food cooked in different ways.
- Consider how you would cook and package this food to sell at affordable prices.

1. Boiling

Boiling is a process of cooking food in liquid over heat. As some nutrients dissolve in liquid, it is good to consume the liquid with the food when this is possible, for example, chicken and vegetable stew.

Cassava and coconut dumplings

Grate 2 large cassava roots.
Combine cassava roots with the grated flesh of 3 coconuts.
Add a dash of salt.
Mix into a doughy substance.
Form balls from the mixture and drop them into boiling water.

Peanut toffee

Put 2 cups of sugar and 1½ cups of water in a saucepan.
Stir over heat until sugar is dissolved.
Bring mix to the boil but do not stir. Let boil until a light golden brown.
Spread peanuts over the base of a dish or oven tray.
Pour hot mix over peanuts.
Let stand until cool and set.
Break into small pieces to eat or sell.

2. Frying

Frying is the process of cooking food in fat or oil over heat.

Deep-frying is when the oil is deep enough to cover the food being fried. Fish and chips are deep-fried. The fish is covered with batter or breadcrumbs before being fried.

Shallow frying is when only a little fat or oil is used. Barbecued sausages or hamburger rissoles are examples of shallow-fried food.

Chips

Peel and cut potato or sweet potato into chips.
Fry in hot oil until crisp and brown.
Sprinkle with salt just before serving.
Estimate a selling price for a cup of chips.

Pancakes

Mix together 1 cup of flour, 1 tablespoon of sugar, 2 eggs and about 1½ cups of milk to make a smooth creamy batter.
Beat well.
Lightly grease a frying pan.
Pour in a little batter and tilt the pan until the batter covers the surface.
When it is dry on top, turn over and cook for 30 seconds on the other side. Remove and place on a plate.
Continue making pancakes until the batter is all used up.
To serve, sprinkle with lemon juice and sugar and roll up.

3. Baking

Baking involves cooking food in an oven. The heat is carried around the oven by air currents. Bread, cakes and biscuits are examples of food cooked in this way.

Baked sago cake

Mix together 1 hand of ripe bananas, 3 cups of grated coconut and 6 cups of sago.
Moisten with coconut water if the mixture is dry.
Make a flat parcel by wrapping the mixture in banana leaves.
Bake in hot stones or hot coals until cooked (about 30 minutes).
This is delicious served in slices with coconut cream.

Banana scones

Sift 2 cups of flour and 2 teaspoons of baking powder into a bowl.
Add ¼ cup sugar.
Rub 3 tablespoons of margarine into the mixture.
Add 2 mashed bananas and mix well.
Stir in enough milk to make soft dough.
Drop spoonfuls of the dough onto a greased oven tray.
Bake in a hot oven for about 10 minutes until cooked.

For you to try

- Try different methods of baking food, traditional and modern.

4. Grilling

Grilling is a method of cooking food from heat that is radiated from hot coals or a glowing metal element. It is a quick cooking method. Toast and sausages are foods cooked in this way.

Scones over hot coals

Rub 6 tablespoons of margarine into 2 cups of flour.
Add enough milk to make soft dough.
Wrap a piece of scone dough around a clean stick.
Hold over hot coals until golden brown.
Drizzle honey or golden syrup over the cooked dough and eat.

Care and Management

HOME, SCHOOL AND COMMUNITY BUILDINGS

1. Houses

Houses are buildings in which people live. Houses meet our need for safety, shelter and loving relationships. They are places for cooking, resting, washing and relaxing.

In rural areas many houses are made from bush materials such as coconut palm, sago palm leaves, pandanus leaves, bamboo, pitpit, kunai and bush timbers. Houses may be built on the ground, on posts, in trees or over water.

In urban areas many houses are made of more permanent materials such as milled timber, concrete blocks, iron and steel. They have a water supply, a power supply, toilets, glass windows, polished floors, painted walls and modern furniture. Some houses in Papua New Guinea are built from a combination of permanent and bush materials.

Care and management of houses varies according to location, size and facilities. It is important that houses are clean, comfortable, attractive, healthy and well-maintained places to live in. All houses need to provide a safe and healthy environment for a family.

For you to try

- Draw a diagram of your family home and add labels. Describe the building materials and list how many people live in your house.
- Imagine building your own house. What kinds of materials would you use for the floor, walls, roof, windows and support posts? Draw a picture of the floor plan and an outside view of your house.

2. Church

Churches are large buildings in which people gather to worship their God. Churches meet our need for religious or spiritual reflection. Churches most often have one very large room with rows of seats for the congregation to sit on and an altar, at the front, where the preacher stands to preach. In towns, most churches are built from modern building materials. Many church buildings in rural communities are built using bush materials. Churches have special areas for communion and baptism. Churches are cared for and managed by the preacher and members of the congregation.

For you to try

- Visit a church in your community and identify the different parts of the building and the building materials used.
-  Compare the design of a church with the design of a family home and explain reasons for similarities and differences.

3. Community hall

Community halls are buildings in which people gather to discuss or participate in activities that are of common interest to members of a community. Community halls meet our need to have a place in which to gather that does not favour any one particular section of a community. Traditionally some villages had a *hauman, hausmrei* or *haus tambaran* for gatherings of men, women or to hold traditional ceremonies. The styles and designs of the community halls and the materials used vary from one society to another.

Haus tambaran, Sepik River

For you to try

- Draw a picture of your community hall.
- Discuss the types of activities people carry out in your community hall.

4. Classrooms and staffroom

Schools throughout Papua New Guinea are made up of many buildings. Classrooms and a staffroom are typical areas in school buildings. Care and management of school buildings is the responsibility of the Board of Management, students and teachers.

Classrooms are places for teaching and learning. Classrooms are equipped with desks, chalkboards, display boards and a teacher's table. Classrooms need to be well lit and well ventilated.

The staffroom is where teachers meet to discuss issues to do with teaching and learning. The staffroom usually has tables, chairs and noticeboards.

For you to try

- Compare and contrast three classrooms in your school. Assess the strengths and weaknesses of each classroom.
- Draw a daily duty roster for the care and management of the inside and outside of your classroom.
- Draw and label a diagram of the school staffroom and explain the purpose of its features.

PD 5. Toilets

Toilets are buildings with special fittings for getting rid of body wastes like urine and faeces. Traditionally in Papua New Guinea, people used bushes, rivers, creeks or the ocean. Today we try to be more hygienic than we were in the past. We have learnt about the importance of hygiene, good health and environmental care.

Key facts

- **Pit toilets** are made by digging a deep pit in the ground and are often used in rural areas. A floor with a central hole is built over the pit. A seat with a lid sits over the central hole. A small house is built around the pit toilet. It is inexpensive to build. Pit toilets can become smelly and attract flies if we don't keep them clean.
- **Toilets built over water** are found in most coastal villages. A walkway on posts is built from the shore for some distance over water. It is important that the walkway is strong and safe for children and elderly people to walk along. There is a hole in the flooring at the end of the walkway. A seat with one or more holes sits over the hole in the floor. A small house is built around the toilet seats. These toilets are low cost to build. The water carries away the body wastes. At low tide, these toilets can cause a bad smell and attract flies.
- **Pan toilets**, also called bucket toilets, are available in some towns. The sanitation department of the town council removes the buckets twice a week and replaces them with clean empty buckets. There is a charge for this service. The bucket sits inside a cabinet. The cabinet has a seat with a lid. They can become smelly and are always built away from the house. The seat, lid and floor need regular cleaning.
- **Septic or sewerage toilets** are found inside houses. They have ceramic bowls with plastic seats and lids. The toilet is flushed with water that comes from a small water tank. The content of the toilet is flushed away along a pipe. Unlike most other types of toilets, they are not smelly. These toilets are easy to keep clean and they are convenient to use, both day and night.

For you to try

- Discuss proper use, care and management of the toilets in your school. Identify problems and suggest solutions.
- Compare and contrast different types of toilets in your community. List the advantages and disadvantages of different types of toilets.
- Identify the cleaning agents and equipment needed to care for the school toilets.

6. Kitchens

Kitchens are places where food is stored, prepared and cooked. Kitchens may be indoor or outdoor. Kitchens are important places because meals need to be hygienic and nutritious for a family's good health. Diarrhoea is a common problem if food is prepared in a kitchen that is not clean. Kitchens are also used for storing and washing equipment used to prepare, cook and eat food.

Kitchens are used more than other rooms in a house and quickly become untidy and dirty. Cleaning, storage and rubbish disposal must occur every time a meal is eaten. All members of a family can share the kitchen tasks.

For you to try

- Think about the care and management necessary for the kitchen in your home.
- Write a list of rules for the care and management of a kitchen.
- Discuss the advantages and disadvantages of indoor and outdoor kitchens.

7. Tool sheds

Tool sheds are small buildings used for storing tools and garden equipment. This equipment could include axes, spades, rakes, bush knives, hammers, saws, screwdrivers, pliers, wheelbarrows and lawnmowers. Sometimes tools are hung on a tool board for neat storage and easy access. Tool sheds need to be kept dry and clean, with tools and equipment stored in a neat and orderly manner. Tool sheds need doors that lock for security. All members of a family can share the tasks.

For you to try

- List things your family has that could be kept in a tool shed.
- Visit a tool shed or garage. Clean it up and remove things not worth keeping.
- Plan a tool shed.

1. Broken walls

Walls offer security and shelter from wind and direct sunlight. Most permanent houses have fibro walls. It is wise to repair a damaged area quickly before it becomes worse and creates a bigger problem.

Repairing broken fibro (external wall)

1. Locate the studs and timber struts (supporting beams) around the broken pieces of fibro. The line of nails shows where to find the studs.
2. Using a straight edge (timber or ruler) and a pencil, mark a line on the fibro that is in the middle of the studs and timber strut.
3. Using the last four teeth of a crosscut saw, scrape out a 'V' cut at least half the thickness of the fibro sheet. Use a piece of timber to guide the saw.
4. Pull the broken section of fibro off the wall.
5. On removing the pieces of fibro, take measurements and transfer these to a new piece of fibro.
6. Cut the new fibro with a fibro cutter.
7. Position the new piece of fibro and nail it with fibro nails.
8. If the required section needs to be waterproof, a piece of galvanised iron cover strip should be positioned on the horizontal joint.
9. Finally, nail the horizontal cover strips. (Use a 2mm drill bit to drill pilot hole.)

2. Repairing internal walls

Most high-cost houses have two walls, the external (outside) wall and the internal (inside) wall. Internal walls improve the appearance of a house, help keep it cool and increase security. There are three main materials used for inside walls: masonite sheets, plywood sheets and fibro sheets. These materials can be bought from all local hardware stores.

Replacing a masonite or plywood wall

1. Remove cover strips from broken walls.
2. Punch the nails with a centre punch.
3. Slowly remove the old sheet.
4. Trace the shape and size of the old sheet onto the new wall.
5. Follow the line with a straight edge.
6. Cut out the waste.
7. Trial assembling the wall to check that the new wall sits flush with the existing wall.
8. Nail the studs and timber struts.
9. Replace the cover strips.
10. Fill the holes with putty and paint in the same colour as the rest of the room.

3. Repairing steps

Steps are important. People walk on them every time they go in and go out of a building. Steps must be checked regularly for loose joins, rot and damage. If you notice that steps are becoming loose, nail the joints and tighten the bolts and nuts. If the steps are rotting, build an extension roof over the steps.

Removing steps and building a new step

1. Remove the steps of the building.
2. Dismantle the timber steps from the two side rails.
3. On removing the timbers, take measurements and transfer them to new pieces of timber. Cut the new timber.
4. Place the new pieces and nail together firmly.
5. If there are only a few steps, nail the new pieces of timber together and then, when complete, join the steps to the building.
6. For a longer set of stairs, firmly join the two side rails to the building. Then join the other pieces of timber to the side rails.
7. Paint the steps to prevent the timber from rotting quickly.

4. Repairing floors

Squeaking floors are a common problem. To get rid of the squeaks:

1. Check if the boards are loose and re-nail them firmly.
2. If several boards are loose, the supporting beam may be sagging (sinking). Restore the joist to its original level using a temporary support and nail a piece of timber, about the same size, to hold the supporting beam.
3. If the boards are badly damaged, cut out the worn section and replace the boards.

Lifting and replacing floorboards

1. If the floorboards are square-edged, insert a support (bolster) near the end of one board. Angle it away from the board and tap it down with a hammer.
2. Raise the end of the board until a claw hammer can be pushed beneath it. Lever the hammer and ease the bolster along the board.
3. When the first board is lifted, use the claw hammer to lift the others.
4. Lever each board gradually and take care not to split the wood.

Fixing creaking boards

1. In the creaking area, punch the nails down so that they sit just below the floorboard surface.
2. Next to each nail head, drill undersize pilot holes for 40mm screws.
3. Countersink each hole and tighten the screw until its head sits just below the surface of the floorboard.

5. Maintenance on tables, chairs, shelves and beds

Over time, tables, chairs, shelves and beds can become damaged. To buy new items of furniture would cost a lot of money and so we need to repair broken and damaged items. Here are some general points to consider:

- Most household furniture is made from wood, plastic or metal. Note the type of material used to make the damaged item.
- Note the style of joints and construction of the item.
- Think of the tools used to make the item and which tools you will need to repair it.
- Assess the damage, and list the materials you will need to collect and use to fix it.
- Plan how to carry out the maintenance or repair.

Furniture repairs

1. For loose joints, apply glue to the joints (PVA glue for wood and contact adhesive glue for plastic).
2. Tightly clamp the joints together and screw or nail the joints.
3. If part of the furniture is broken and needs replacing, remove the broken piece.
4. Measure the broken part, transfer the measurement and cut out a replacement piece.
5. Nail or screw the new piece into position.
6. Sandpaper and wipe clean.
7. If the legs are metal, thoroughly brush and sandpaper any rusty areas.
8. Apply paint (colour to match the rest of the item).
9. Leave it to dry under a shady tree, away from direct sunlight and dust.

6. Leaking taps, broken water pipes

Neglecting simple faults, like leaking taps, creates problems that require major work. This can cost a lot of money and water is wasted. It's simple. You can learn to put a new washer on a dripping tap. Taps come in different shapes but the inner structure is basically the same. When the washer becomes worn, the tap drips.

Leaking taps, showers and toilets

1. Shut off the main tap to prevent water from flowing.
2. Loosen the hexagonal head with an adjustable spanner.
3. Unscrew and loosen the nut by hand.
4. Lift the top off the tap and put it to one side.
5. On most taps, the jumper valve fits loosely and simply pulls out.
6. Grip the edge of the jumper valve with pliers.
7. Undo the nut and take off the old washer and check the valve seat inside the tap.
8. Fit the new washer and replace the metal washer and nut, if the thread is damaged, get a completely new valve.
9. Put back (reassemble) the parts.
10. If the washers need frequent replacement, the valve seat may be damaged. If this is the case, remove the whole tap and replace with a new tap.

Repairing broken water pipes

1. Turn off the main tap to prevent water from flowing.
2. Cut the broken section of the pipe with a hacksaw.
3. Transfer the measurement to a new pipe and cut it.
4. Apply pipe glue onto both ends of the new pipe.
5. Fit the new section so that it sits outside the top pipe and inside the bottom pipe.
6. Leave the pipe joints to dry completely.
7. Turn on the water to see if there is any leakage from the joints.
8. If there is water leaking from the joints, use plumber's sticky tape to wrap around the thread of the joints.

OTHER IMPROVEMENTS

In addition to building maintenance and repairs, we need to make sure that buildings and grounds look clean, tidy and attractive. Here are ways to make these improvements.

1. Repainting walls

Fresh paint makes a wall look more attractive. Paint also protects the wall surface from wear and tear, so that it lasts longer.

Painting walls

1. Use an abrasive paper to scrape out dirty spots and smooth old paint.
2. Use wood filler or putty to fill in cracks in the wall.
3. Wipe off any dust or dirt with a cloth.
4. Apply undercoat paint to the wall and let it dry.
5. Apply gloss paint to make the external walls look attractive.
6. For internal walls, apply semi-gloss paint. Semi-gloss paint reflects less and is gentler on your eyes.

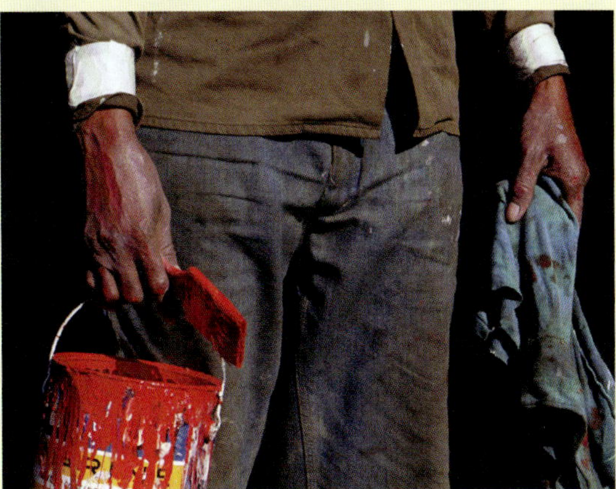

2. Replacing louvres or window frames

In the past, many houses were built without windows. In many areas this was done for security, and in the highlands this was done for warmth. Today people know that it is important to have windows to let in fresh air and sunshine. In a modern house, louvre windows are common. Louvres are made from metal and wood.

Replacing louvres and window frames

1. Remove the curtains or tie them back if they get in the way.
2. Open the window so that the louvres are in a vertical position and remove the broken louvres.
3. On removing the damaged or broken louvres, transfer the measurements and cut out a new louvre or window frame.
4. Place the new pieces into the louvre frames and screw the window frames until firm.

3. Landscaping

Landscaping is making improvements to the grounds surrounding your home. A well-planned outdoor living area with a boundary hedge or fence, lawns, flower beds and shady trees looks attractive and gives your family the opportunity to spend pleasant leisure time outdoors. Here are some landscape ideas for around your home:

- **Setting boundaries.** In rural areas, bush timber can be used for fencing. In urban areas, high wire fences are a common sight. Hibiscus or tanget can be planted in rows to form a hedge.
- **Making flower gardens.** Flower gardens are usually planted close to the house and along the edges of paths. These add beauty to your grounds.
- **Setting lawns.** Lawns are grass-covered areas. They should be cut short regularly. Rake dead leaves and gather rubbish to keep lawns neat and tidy.
- **Constructing pathways.** A path leading to the front door is usual and a path from the back door to the clothesline, and to other garden areas, is useful.

Landscaping around buildings in your school

1. Begin by setting the boundaries around your school. Erect fences and plant trees or plants to indicate the boundary.
2. Make pathways to the classrooms and other buildings in the school. The pathway can be surfaced using concrete, gravel or large flat stones.
3. Create lawn areas for sports, assemblies or recreation. Regularly rake leaves, cut the lawn and trim edges to keep the lawn areas neat.
4. Plant a few shady trees for people to rest under shade.
5. Make flower beds and plant attractive flowering plants. Line stones along the edges of flowerbeds or fence them in an attractive way.

4. Building extensions

Extending a building means to add something on to the building. This could be a roof overhang, veranda, garage or room. Extensions to buildings are made when a family needs extra space for living. A simple extension could involve constructing a roof over external steps to prevent them rotting in wet weather. A veranda would be a more complex extension.

For you to try

- Go for an excursion. Look at buildings and study extensions that have been made. Draw pictures of an extension to a building. Make a list of materials needed and estimate the cost. Discuss the skills needed to construct the extension.
- If possible, plan and build an extension onto a school building. If you are not confident, ask for help from a skilled and experienced person.

5. Constructing footpaths

Footpaths provide people with firm ground and direction to get from one place to another. Footpaths can be made from soil, but they may become muddy in wet weather. Footpaths can be surfaced with concrete, gravel or flat stones.

For you to try

- Plan a pathway for your school grounds. Begin by observing tracks to see where people walk to get from one place to another. Talk to the head teacher to get his or her support. Construct the pathway. It can be surfaced using concrete, gravel or flat stones.

6. Botanical garden

Papua New Guinea is a beautiful country with thousands of varieties of beautiful plants. A botanical garden could be built in an area of your school to show off the many interesting trees and plants that grow in your local area. You can earn an income by selling cut flowers, plants in pots or plant cuttings. You could even charge a small fee for visitors to see your botanical gardens.

For you to try

- Plan to establish a botanical garden in an area of your school grounds. Investigate trees and flowering and non-flowering plants that grow in your locality. Choose a suitable area. Explore different options and design the garden. Develop and record your plan. Put your plan into action and evaluate the outcome.
- On a smaller scale, you could turn your classroom garden into a botanical garden. Consider selling cut flowers, plants in pots or plant cuttings.

AREAS TO ASSESS FOR MAINTENANCE

1. Building improvements and maintenance

Buildings and grounds are essential for our shelter and comfort. It is important that buildings and their surrounds are attractive, well maintained and long lasting. Care and maintenance of buildings should be everyone's responsibility. It is better (and costs less) to repair faults at an early stage. If left, a small repair can become a large and expensive problem to fix.

For you to try

- In small groups, walk around the school and check school buildings to find areas that need work.
    1. Plan how you will carry out the repairs. Consider the method, tools and materials required and also the cost.
    2. If the cost of the materials is high, talk plans over with your teacher and raise the matter with your head teacher and the school board of management who fund repairs.
    3. Following your plan, carry out the work required.
- Landscape the school grounds so that they look beautiful.
- Carry out maintenance on your family home too.

2. Parts of buildings to check

Parts of buildings that need regular checks for maintenance and improvement include doors, flooring, steps, cupboards, roofing, gutters, tanks, bathrooms, kitchen, living room, garden, bedroom, veranda and railings. Anyone can make repairs and home improvements by following simple 'do it yourself' procedures.

Repairing a leaking roof

When it rains, a wet spot on your ceiling is usually a sign that there is a leak in the roof. Check where the water is coming in. Most roof leaks are caused by loose roof nails, rusty holes or by damage to the overlap of corrugated iron sheets.

Steps to follow:

1. If the roofing nails come loose, try larger or new roofing nails.
2. Hammer each nail in a new place.
3. Patch the old nail holes using a thick coat of bituminous roofing compound (tar) and place the patch into position.
4. Rust holes in a corrugated roof can be patched in this way too.
5. For leaks from roofing nail washers, replace with a new roofing nail or patch it with the bituminous compound.

Repairing roof guttering

Proper care of guttering makes your house, gutters and downpipes last longer. There are three ways to do this:

1. Scrape out dead leaves from gutters every three months. They hold water, and they can clog downpipes and stop the flow of water. If you have bituminous paint, coat the inside of galvanised guttering.
2. Remove the old gutter and replace it with new guttering. If you find that only one or two sections of the guttering have rusted through, cut away the damaged section with a hacksaw before loosening nails and brackets.
3. Nail the joined guttering firmly onto the fascia board.

Repairing downpipes

1. Probe down the pipe with a long stick to clear anything clogging the downpipe.
2. Cover the upper ends of downpipes with mesh wire to prevent clogging.
3. Remove the whole pipe from the wall to replace a broken section of downpipe.
4. Cut the damaged section of the pipe with a hacksaw and tin-snips.
5. Fit the new section so that it sits outside the top and inside the bottom pipes. To do this, cut along the seams of the pipe.

Wise Consumer

CONSUMER RIGHTS AND RESPONSIBILITIES

1. Consumer rights provide a consumer with guidelines for consumer protection and form the basis of consumer laws

The government has set laws to protect consumers. The government has a Goods Act. Items sold in shops must be in good condition. Shops should not sell food that has passed its use-by date. There are health regulations that require shops to be clean. Health inspectors can fine or close shops if hygiene standards are not met. This is especially important for shops that sell cooked food. The government has established a Consumer Bureau. It monitors prices and inflation rates and handles complaints.

For you to try

- Interview people in your community and record information about the types of problems faced by consumers in Papua New Guinea.
- Write to your nearest Consumer Affairs Council and ask for information about the work they do. You can find their address in a telephone book.

2. Consumer rights

We are all consumers and we have rights. We have a right to:

- **Safety.** Goods and services we pay for should not endanger us in any way; for example, the food we buy should not give us food poisoning.
- **Honest information.** We must be given reliable, honest information about goods and services that are advertised and that we buy. For example, sale prices must be genuine reductions.
- **Choice.** We should be given a choice of goods and services. This avoids one business controlling the quality and charge of service. For example, we should have a choice of banks, health services and food products.
- **Be heard.** If we have problems with the quality or charge for goods and services, we have the right to be heard. For example, we have the right to complain if we are given incorrect change at a shop.
- **Redress.** If there is a problem, it should be solved. For example, a faulty item should be replaced or money returned.
- **A healthy environment**. We should have access to a clean water supply and rubbish collection, for example.
- **Consumer education**. We should learn how to use money wisely to get value for the money we spend.

Rubbish bin

For you to try

- Talk about consumer problems. Choose an issue that concerns you and write about it.

3. Consumers are responsible for obtaining information and taking appropriate action

As a consumer it is important to learn how to be wise in using your money. You are the best person to help yourself. You are responsible for obtaining information and taking appropriate action.

Imagine that you will buy a kerosene lantern. You should ask questions about how to raise the glass, raise and lower the wick and where to put the kerosene. You should check that it is well built with no signs of rust or holes where kerosene could leak out.

Imagine that you will buy a tin of powdered milk. You need to read the instructions on the packet to know how much milk powder to mix with a certain amount of water to get full strength milk.

For you to try

- Read the instructions on packets (for example, 2-minute noodles) or tins that you buy.
- Explain to someone who cannot read how the product should be used.
- Discuss the importance of being able to read.

4. Consumers are responsible for critical awareness, taking action, social concern, environmental awareness, working together

Consumers need to be responsible in spending their money and consumers should expect to receive quality goods and services.

Consumers need to be critically aware of prices in shops. Sometimes the price of a product is higher in one store than it is in another. Sometimes you save money by buying a large packet of something instead of two smaller packets.

Consumers should show social concern over poor delivery of services such as power, water, health, education, and transport or rubbish collection. You have a right to take action and demand quality service.

Consumers should work together to take action if they share a common problem.

- If transport is a problem, what can be done about it?
- If there are long lines to get money at the bank, what can be done about it?
- If opening hours of stores are a problem, what can be done about it?

For you to try

- Form an action group and talk about a common consumer problem. Make a responsible and positive plan to do something about the problem.

QUALITIES OF A WISE CONSUMER

1. Comparing prices of goods and services

A wise consumer is always aware of prices for common goods they buy. Often there is a difference between prices in stores. For example, one shop may sell a jar of coffee for K5.99 while another shop sells the same brand and jar-size of coffee for K4.99. A wise shopper can save money by buying items at the lowest price.

Even at the market, people sell their food at different prices. Someone may sell a hand of bananas for K3.00 while another person may have a similar hand of bananas for only K2.00.

In the same way, similar services may have different prices. Someone may charge K20.00 to cut the grass while another person may charge only K15.00 to cut the same area of grass. Someone may charge K2.00 to wash a vehicle while another may charge K3.00.

For you to try

- Compare prices of goods and services in your local area. Become aware of where to shop to get the lowest price for common goods and services.
- Cut out advertisements in newspapers and identify different prices for the same items.

2. Checking the quality of different brands

Cheapest is not always best. There is a wide range of prices for clothes, tools and household items. A wise consumer checks out the quality of different brands before deciding to buy. It is wise to buy the best quality that you can afford.

A T-shirt may be cheap but if it stretches out of shape quickly, then it is not good value for money. An axe may be cheap, but if the head falls off the handle after a short time, then it is not good value for money. Cutlery may be cheap but if soon after purchase a fork bends out of shape, it is not good value for money.

For you to try

- Name brands of tinned fish, tinned meat, rice or hard biscuits that you think are better than others. Give reasons for your answers.
- Give examples of cheap products you or your family have bought that were not good value for money. Describe what happened to make you disappointed in your purchase.

3. Writing a shopping list

A wise consumer always makes a general shopping list before going to the shops. In this way needs are met and impulse shopping is avoided. (Impulse shopping is when you are tempted and buy something without thinking carefully about whether you need, or can afford, the item.)

A shopping list may include food from the market, food from the shop and household items. From time to time special items, like stationery, clothes or new cooking pots, may be needed.

For you to try

- Make a general shopping list for your family's weekly shop. Estimate the total cost of the items on your list.

4. Observing consumer laws for protection

Many perishable goods are required by law to have a 'use-by date' on them. A wise consumer always looks for the use-by date on products and does not buy products that are past the use-by date. This law is there to protect the consumer.

Many electrical products have a warranty period. If something goes wrong within the warranty period, the store should fix the product for free. A wise consumer keeps receipts and warranty cards in case something goes wrong with the product they have purchased.

Cigarettes by law are required to carry a message that warns that smoking is dangerous for your health. Other products may carry warnings to protect consumers. Cleaning liquids may show 'Dangerous when swallowed'.

For you to try

- Look for use-by dates on products. In your own words, explain what the label means.
- Look for warning signs on products. Explain how they protect consumers.

5. Keeping records, such as receipt of purchase, in a safe and clean place for future reference

Most big shops give a receipt for goods purchased. The receipt shows where something was bought, the date it was bought, what was bought and how much was charged. A wise consumer always checks the receipt and also checks that the correct change is given. Sometimes receipts are hand-written and sometimes they are printed by machine.

It is always wise to keep receipts for expensive items such as sewing machines, electrical equipment, lawn mowers or generators. If there is a problem and you return an item to a store, they will require proof of payment. It is important to be able to give them a receipt. A wise consumer keeps records, such as receipt of purchase, in a safe and clean place for future reference.

For you to try

- Draw a receipt form and fill in the details of something you would like to buy.
- Collect and make a display of machine-generated and hand-written receipts. Explain the information given on the receipt to another person.

- Design and make a folder or file box for your parents to keep their receipts. Consider how these folders or boxes could be sold to the general public.

Making Things

ITEMS OF BENEFIT TO INDIVIDUALS AND COMMUNITIES

Many Papua New Guineans make things for personal use, to give to others, to improve the community or to sell. The things we make may benefit individuals and communities. They may be small things (a photo frame or candleholder) or something large (a drum oven or a new footpath through a village).

Things to consider when planning projects are the availability of resources, time, skills, expertise, costs and space.

- What resources are available from which to make things?
- What materials and equipment are required?
- How much time is needed to do the task?
- When is time available?
- What skills or techniques are needed to make the item?
- Do we have the expertise? If not, who can teach us?
- How much will an item cost to make?
- From where will the funding come?
- How much space is needed?
- Where is there suitable working space?
- How can materials and equipment be stored safely?

For you to try

- Talk about the availability of resources, time, skills, expertise, money and space for projects you could undertake. Be as imaginative as possible. Visit different parts of your community looking for ideas. Be aware of opportunities to make things that would benefit individuals or communities.

STARTING PLANS

1. Generating ideas through brainstorming

Brainstorming is a way of getting ideas from as many people as possible. Everyone can suggest ideas, and all ideas should be accepted without judgment. Write all the ideas on a large sheet of paper. Look at the ideas and then see if you can add some more.

- What household items could be sewn? Curtains, pillowcases, bed covers, cushion covers or storage bags?
- What clothing could be made? Dresses, skirts, shirts, shorts or trousers for children or adults?
- What craft items could be made? Stuffed toys, baskets, mats, photo frames, beads, wall decorations, plant holders, cane curtains, hangers or carvings?
- What things could be made from wood? Coconut scraper, storage boxes or furniture such as beds, chairs, a table, cupboards or bookshelves?

- What things could be made from metal? Chairs, beds, lamp stands, coconut scrapers, barbecue plates, drum ovens, strainers, rakes, water tanks, buckets or tin trunks?

For you to try

- Brainstorm ideas of things you could make for personal use, to give to others, to benefit the community or to sell. Get ideas from different people – the head teacher, community members and children from other classes. Visit homes and markets and write down ideas for hand-made items. Allow a week to build up a bank of ideas of things to make from available resources.

2. Looking at samples and adapting them

Look at samples of handmade items and consider how they could be adapted to make something better or more interesting. This is called being **innovative**. This means to change something by bringing in a new idea.

- Instead of a plain bamboo flower holder you could create a design or add pig's tusks or seeds.
- Instead of a plain coconut bowl you could line it with fabric, cut different shapes into the shell, leave some husk on the shell or make a macramé string holder for it.
- Instead of a plain bead armband you could work a pattern, or someone's name, into the design.
- Instead of a plain fabric pillowcase, you could embroider, tie-dye or appliqué a design.

For you to try

- Draw a diagram of a tin or bottle that could be used as a flower vase. Then describe, using diagrams, three different ways it could be made more attractive or interesting.
- Be imaginative and think of different ways to adapt a handmade item to make it different and more interesting. Use diagrams to illustrate your ideas.

3. Creating your own designs

Creating your own designs is a way of using your imagination to make things that no one else has made.

Some Papua New Guineans carve wonderful frames for wall clocks. They may carve crocodiles or birds of paradise in the frame around the clock. This is not a traditional handmade item. These Papua New Guineans have created their own designs.

For you to try

- Create your own design for a table setting that includes a place mat, eating bowl or plate, knife, fork, spoon and drinking cup. Describe the materials you could use.

4. Imitating

Another way of developing ideas is to imitate or copy something that you have seen.

As you visit different homes, for example, you may see something and think, 'That would be useful in my own home.' Always be looking for new ideas to copy and make your own. It could be a sawdust stove, a fold-up table or chair, or an idea for an outdoor seat or barbecue stand.

For you to try

- Think of metal, wood, craft or sewn items that you have seen and that you think you could copy.
- Tell stories about where you went, what you saw, how it was made and describe its usefulness and why you feel you could copy it.
- Draw labelled pictures of your ideas.

5. Choosing the best idea

Think about all the ideas that you have listed and discussed.

- Consider the availability of resources, time, skills, expertise, cost and space.
- Make a short list of projects you could do.
- Identify the most appropriate idea to decide what you will make.
- Draw pictures to explain the design and production steps.

6. Developing a plan

By now you should have chosen something to make. The next step is to develop a plan of action.

- What will the product look like?
- What size will it be? What resources are needed?
- Where will resources come from?
- What needs to be bought and can you afford it?
- What are the steps involved in making the item?

List activities in a sensible order.

Action plan to make a bamboo cassette holder

1. Draw a design for a bamboo cassette holder showing measurement details.
2. Get the tools and equipment needed for the project, for example: saw, pencil, ruler, sandpaper, glue, varnish, and brush. Borrow or buy if necessary.
3. Find and cut bamboo of a suitable size and length.
4. Use a pencil and ruler to mark where the bamboo pieces should be sawn at regular intervals. Use a cassette to measure the width needed for the cuts.
5. Saw and remove pieces from the bamboo where cassettes are to be stored.
6. Cut pieces of bamboo for the legs or stand.
7. Use sandpaper to smooth bamboo surfaces and cut edges.
8. Glue legs to the bamboo rack.
9. Varnish and let stand to dry.
10. Display for sale with price tag and some cassettes.

For you to try

- Write down a plan of action for the object you are going to make. This will probably be a procedural text with instructions to be followed. Start by drawing a picture of the thing you will make and end with an item ready to be sold.

7. Deciding on materials required

What you are making will determine the materials you need to collect or buy. Materials need to be carefully selected to suit the needs of your design. Consider strength, flexibility, texture, shape and colour.

Items available in your environment could include seeds, sheets, bamboo, coconut, pandanus, bush fibres or items suitable for recycling. Items to be bought could include timber, fabric, nails, screws, sandpaper, varnish, paint and thread. As you list the things you need, also list the size, length or quantity you need for your project.

You also need to think about tools and equipment. This could include hammers, saws, paintbrushes, scissors or needles.

For you to try

- Write down a list of consumable items you need to collect or buy and the quantity of each that you need. This will probably look like a shopping list.
- Then list the tools or equipment that you will need. Consider what is available and where to get the items that you do not have.

8. Researching and investigating design, materials, skills required, tools and equipment

Techniques required for **sewing** include knowledge about fabric, pattern making, measuring, cutting, stitches and seams.

For **woodwork** you require skills that will enable you to choose suitable timber, measure, saw, join, nail, hammer, screw, drill, chisel, sandpaper, polish or paint.

For **metalwork** you require skills that will enable you to choose the most suitable metal, measure, cut, grind, solder, drill and paint.

For **craftwork** you need skills that enable you to cut, sew, weave, thread, carve, glue or varnish.

It is important to learn the skills and techniques needed to work with materials in ways that suit the challenge of your construction. Resources should be used economically and equipment should be used safely.

For you to try

- Discuss the skills, techniques and construction processes needed to realise your design.
- Use scrap material, or make samples, to demonstrate and practise skills.

9. Collect and gather resources

Collect and gather all the resources that you will need before you begin. Check that your materials are in good condition. Ensure that you have a workspace that is clean, well lit and ventilated with enough space for easy movement. Arrange resources so that they are ready for when you need them.

Plan your time so that resources are collected and gathered for when they are needed. Make sure that you get good value for any money that you spend. Compare both price and quality of items at various stores before buying them.

For you to try

- Plan in advance the time needed to collect and gather resources. Set a date before which all resources should be assembled ready to start the project. Collect and gather resources by the set date.

10. Production and manufacture

When you have your action plan and materials in place, you are ready to start production and manufacturing. Simply do it. Have a go. Take the risk. Be adventurous!

Follow your plan. Use equipment safely. Apply appropriate production techniques. Manage time and resources effectively. Monitor and control the quality of the product or service. Solve problems as they arise and adjust plans. Keep going until the task is complete. The aim is to produce a quality product that will appeal to your customers.

11. Testing and evaluation

It is important to check that the item you have made works in the way you imagined. Try your products out. Look for faults or mistakes. Answer these questions:

- Was the design successful? Why? Why not?
- Which features are successful?
- Which features don't work?
- What were the frustrations in making the item?
- How could the product be improved?
- Would you repeat the process to make the same thing again? Why? Why not?
- What changes could be made?
- Will you try a different project?

If you plan to sell the products made, calculate the cost of production and add a reasonable, but not excessive, profit margin to determine the selling price. Consider:

- When and where will the product be sold?
- How will the product get to where it is to be sold?
- What will happen to unsold items?
- Will it be advertised?
- How will people know that it is available?
- Who will sell it?
- Will money be needed to give change to buyers?
- Is a container needed to put the money in?
- What records need to be kept?

Keep records of sales. Use competitive marketing strategies. Advertise products in various ways. Identify good locations for selling a product, techniques to attract customers, and appropriate ways to respond to customers. Recognise the importance of the customer and his or her right to be satisfied with the product purchased. Appreciate the value of a polite and friendly manner in dealing with customers.

For you to try

- Reflect on your initial intentions and plans.
- Did the products and processes meet your expectations?
- Identify successes, failures and constraints.
- Consider the viability of the project and decide whether to continue with it, continue with modifications or undertake an alternative project. Develop criteria to assess the success of a project.

Community Development

Chapter Summary

In this chapter you will have an opportunity to:

- ✔ Reflect upon individual strengths and capabilities and how to use them within a community in a positive way.
- ✔ Identify and establish networks within a wider community to promote effective access to information to benefit the wider community.
- ✔ Plan cooperative projects that require community and school participation.

Syllabus References

Strand: Community Development

Substrand 1: Knowing Communities

Outcome: 7.3.1 Reflect upon personal strengths and capabilities and consider how they might use these to contribute in a positive way within the community.

Substrand 2: Communication

Outcome: 7.3.2 Identify and establish network partners within the wider community to promote more effective access to communication.

Substrand 3: Community Projects

Outcome: 7.3.3 Initiate and plan cooperative projects that encourage community and school participation.

Knowing your Community

PERSONAL STRENGTHS AND CAPABILITIES

1. Communication skills

Communication is about passing and receiving information. We all have to communicate with other people to pass on and receive information. Learning to communicate well with others is an important social skill that can be learnt.

Communication can be oral or written. In oral communication, we should speak clearly and in a way that is easy for listeners to understand. If we are listening, we should listen with interest, to understand the message. Listening is important for good communication.

In written communication we should write clearly and in a way that makes it easy for the reader to understand. If we are reading, we should read with interest to try and understand the writer's message. Good communication promotes good understanding.

For you to try

- In threes, practise the skill of speaking clearly, listening with interest and writing well. The speaker can talk about an important issue, like the importance of living in a healthy environment. The listener should listen with interest and the recorder should write down the important points made by the speaker.
- After five minutes, the recorder writes down the main idea. The speaker can correct any wrong information. Each student should have a turn as speaker, listener and recorder.

2. Leadership PD

Every day we make decisions about our own lives. The leaders in our community make many decisions for us. Leaders are a special group because they have the power to influence or control the lives of other people. They often take decisions that affect others in the community.

We all belong to groups with different levels of leadership. We may have a high position and a low position in another place. For example, a school principal can have a high position in a local community. He or she can make decisions that affect children, parents and teachers in the community, but the same principal may have a low position within the Ministry of Education and may not have any influence on important decisions that affect schools throughout the country.

In traditional Papua New Guinean communities, everyone knew their leaders and the leaders knew the people they were responsible for. In modern Papua New Guinea, there are leaders who make decisions that affect people throughout the country. But many people will never meet these leaders, because it is impossible in large societies for leaders to be known in the way leaders were known in smaller traditional communities.

People can become leaders in many different ways:

- **Hereditary leaders** may inherit their positions of leadership from their families.
- **Appointed leaders** may be appointed because of their qualifications and experience.
- **Elected leaders** may be chosen by voting, for example, a class captain or a Member of Parliament.
- **Leaders by force** may be obeyed because they are feared.
- **Leaders because of wealth or position** – wealth may give leaders power over people who depend on them.
- **Leaders because of skills or knowledge** – special skills and knowledge can give leaders power to influence people such as sport captains, magicians and rainmakers.

Leadership is an essential role in our society. We need leaders to make decisions on our behalf. Leaders should be:

- People who are respected in the community.
- People who are elected or appointed to an important office.
- People who are able to make wise decisions about important public issues.

A leader in modern society has many responsibilities. A leader's job is to serve the people and to make decisions in the best interests of most people.

However, everyone can be a leader. In small informal groups you can have leaders who lead because:

- They have a special talent or gift (public speaker, team captain…).
- They are talkative and a good listener (school captain…).
- They are physically bigger than others (front-row rugby player…).
- They are popular (entertainers…).

For you to try

- Invite a community leader, such as a church pastor, local councillor or head teacher, to tell how he or she became a leader.
- Make two lists. In the first list, write the qualities of a good student leader in your school. In the second list, write the leadership skills that you have. Compare the two lists and discuss the differences.

3. Public relations

Public relations refers to the way in which people in the community find out about community services and businesses and organisations.

Many people and organisations (like companies) take care to improve their public relations, so that people in general will think well of them. They may sponsor a sporting event, for example, or they may donate money to a charity. By making sure they have good public relations, they expect that people will support them or their activities.

Public relations is about creating and maintaining goodwill and a good public image. Good public relations can be important for individuals and groups of individuals in companies or other organisations, especially when they want to influence the community. Being friendly and helpful develops good public relations with others. People will then tend to support these group's activities.

For you to try

- Visit your local store and observe how people are treated by security guards and shop attendants. List three examples of good public relations.
- Describe examples of good public relations you might expect from a bank, or Air Nuigini, or in a store.

- Look through newspapers and magazines and cut out sections where companies and organisations advertise and promise their products and services. How do these advertisements promise good public relations?

4. Skills

Sporting capabilities PD

When we develop our sporting capabilities and our physical skills, we help our community and ourselves. Taking part in individual sporting activities such as swimming and running, and team sports such as rugby and basketball, will benefit our bodies and minds, and sport helps us relate to other people within our community.

The major benefits of developing our sporting capabilities include:

Physical benefits

- Increased energy levels
- Weight control
- Improved personal appearance
- Good muscle development
- Improved coordination and flexibility
- Fitness.

Mental benefits

- Increased self-confidence
- Improved concentration and mental alertness
- Increased ability to relax.

Social benefits

- Meeting new people
- Improved friendships
- Learning to work as part of a team.

Class Project 

List all the types of exercise in which you have participated over the past week. Include activities carried out in your home, as well as sporting activities. Draw up a weekly exercise grid (as below) and next to each activity:

- Write the amount of time you have spent on this activity.
- List the intensity level of the activity.
  LOW = required little effort e.g. leisurely walk.
  MEDIUM = more energy required e.g. walking briskly, dancing.
  HIGH = heart rate and breathing increased greatly e.g. running, swimming.

Weekly Exercise Grid – Example

Monday	Tuesday	Wednesday	Thursday	Friday	Saturday	Sunday
Walked to school (30 mins) LOW	Walked to school (30 mins) LOW	Walked to school (30 mins) LOW	Walked to school (30 mins) LOW	Walked to school (30 mins) LOW	Worked in the garden 180 mins HIGH	
Class physical education lesson (50 mins) MEDIUM	Went swimming after school (60 mins) MEDIUM	Rugby training (60 mins) HIGH				

- Once you have completed your weekly exercise grid, write a summary of your physical activity.
- Using all student information, make a grade fitness profile. How fit are the members of your class? Plan a physical and sporting program to increase the fitness level of all class members.

Technical (practical) skills

The development of a community depends on having people in the community with good technical skills.

Think about your community and the people you know who have good technical skills – they may be good at gardening, bilum making, house building, carving, nursing, teaching, fixing or making things. Think about the benefits that these people bring to the community.

You may be good at carpentry, cooking, sewing, craft work, carving or canoeing. As you grow older you might learn other technical skills such as driving a car or training to be a doctor, nurse, teacher, mechanic, electrician, plumber or a computer technician. It is important to have long-term and short-term goals about the technical skills you would like to develop to benefit yourself, your family and your community.

For you to try

- Invite a skilled person to school to demonstrate their skills such as weaving a basket, weaving a bilum or making a fishing spear.
- Make a simple item using the above skills.

Interpersonal skills PD

We all need to be involved in relationships with other people. We form relationships because we have certain basic needs to be satisfied. The development of good relationships depends on the development of good interpersonal skills. Interpersonal skills are the social skills we need to relate well to other people in our community.

Interpersonal skills include:

- effective communication skills
- decision-making skills
- cooperation skills
- listening skills.

Good interpersonal skills depend on the development of good personal qualities, such as:

- developing trust
- being considerate and polite
- honesty
- loyalty
- considering the feelings of others.

These qualities and skills help the members of a group to work well together to achieve common goals.

For you to try

- Test your group's interpersonal skills:
  1. Form groups of six to eight students and stand in a circle. Connect hands, making sure you do not connect hands with a group member directly next to you or that you do not join both hands to the same person.
  2. The group should now look like a large knot.
  3. The aim of the activity is to unravel the knot, without letting go of any of the hands, by working together to solve the problem.
- Think of someone you regard as a really good friend. Identify why it is important for that person to be your friend. List the strategies you use to maintain the friendship. What are your strengths? What are the constraints?

Academic skills

You gain academic skills at school. You learn to read and write and gain knowledge about social science, language and the arts. You obtain useful skills through **Making a Living**. You learn the importance of being a good community member.

In various ways your teacher will assess your work during the year, and there will be times when you have to study for tests and exams. When you join the workforce you may again have to study for tests and exams if you are training for particular jobs, for example, as a policeman or a nurse, or if you are trying to gain extra skills.

Studying well

The key to studying well is to be organised. The following steps will help you plan and organise your study program:

1. Organise your study material. Make sure that your work is complete and up to date.
2. Make sure you know what kind of assessment task you are going to do. Then practise the skills you need to complete set tasks.
3. Use a variety of study methods – such as memorising facts, drawing diagrams, summarising information under key ideas.
4. There are always learning areas that we don't like or find hard to do. The more time you spend on your weak learning areas, the stronger you will become in them.

In time you will be able to use your academic skills to help yourself and your family and contribute to community development in practical ways. You could use your academic skills to help the community to write a project proposal for community development or to write a letter to a government official or NGO group to raise awareness on important local issues.

For you to try

- Plan a career day and invite special people (such as a police officer, doctor, businessman or nurse) to tell what skills and education level they needed to get their jobs.

Problem-solving skills

There can be many problems in life, but a problem can be seen as a challenge. Problems are like waves in the ocean. They come at different intervals and in different strengths. Everybody faces problems in life and everybody needs a positive attitude to find ways to solve problems.

How to solve a problem:

- Identify the problem you want to solve.
- Find out as much information as you can about the problem.
- Identify the difficulties you think you are going to have.
- Consider the courses of action you can take to solve the problem.
- Consider all possible solutions.
- Choose the best solutions with reasons.
- Try out the solutions you think will work.
- Evaluate the solutions.
- Choose the best solution and put it into practice.
- Think about the solution you have chosen. Could you improve it?

Class Project

Identify an environmental health problem in your area. Follow the above problem-solving guidelines to help you find a solution to the problem.

Record the process you follow under these headings:

- Identifying the problem
- Finding information
- Identifying difficulties
- Suggesting and trying different solutions
- Choosing and implementing the best solution
- Evaluating the solution
- Improving the solution.

For you to try

- What would you do if:
  1. *Raskols* held you up with a gun?
  2. Fire broke out in your house?
  3. There was a strong earth tremor in the next few seconds?
  4. A fight broke out in the community?
  5. You did not have the opportunity to go to secondary school?
- Write down five situations that could present problems in your life. For each problem, write two options of actions you could take to solve the problem. Discuss your preferred options with a partner.

5. Personal strengths

We all have different qualities that make us unique. We are all good at some things and not so good at others. The things we are good at are our strengths. The things we are not so good at are our weaknesses. We all need to look at ourselves to know our personal strengths and weaknesses. Our strengths and weaknesses don't remain the same all our life.

Mirou has identified some of her strengths and weaknesses in the list below:

Things I like:

Playing volleyball with my friends
Swimming
Eating hamburgers

Things I don't like:

Working in the garden
Getting up early for school
Sharing with my brother

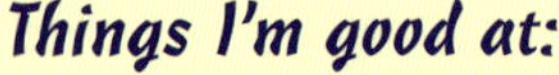

Things I'm good at:

Not giving up when I can't get something right
Writing stories

Things I need help with:

Maths problems
Sticking by my friends
Not losing my temper

For you to try

Work with a partner to make a list like Mirou's.

- Write down the things you like and don't like.
- Write down the things you are good at and the things you need help with.
- Do the same for your partner.
- Did your partner identify the same likes and dislikes, strengths and weaknesses for you as you did for yourself?

A goal is something that we try to achieve within a certain period of time. Read about Fabian's goals in the story below:

Fabian's goals

Fabian is a good swimmer. He does well in school swimming events but he hasn't been successful in making the provincial swimming team. Fabian practises every day so that he can make the provincial team. To do this, he has to get up early so that he can swim before school begins. Three afternoons a week, he goes to the public pool for group coaching. On weekends, he often takes part in swimming events.

'I've found that I have to be very organised and disciplined,' said Fabian. 'I have to make time to do my school work and chores at home and also have fun with my friends. I must also eat healthy food and get enough sleep because swimming takes a lot of energy. Before I set my goal to swim for my province, I found it difficult to train hard,' continued Fabian. 'Now I'm more focused. I know I can't be a swimmer all my life but I know I want to do something connected with sport, such as being a personal trainer or a coach.'

- What special talents does Fabian have?
- What is Fabian's short-term goal? What is his long-term goal?
- Lists the strengths that Fabian has developed in order to achieve his goals.

For you to try

Fabian's special talent is swimming. You may be talented in certain sports or skills such as crafts, painting, sculpture, and may want to improve, like Fabian.

Try this activity to identify your own talents.

- Identify a special talent that you have.
- List the ways in which you can develop your talent.

For you to try cont.

- Set yourself specific goals for developing your talent. List what you would like to achieve by:
  - The end of this year.
  - The end of next year.
  - By the time you have finished school.
- Identify the strengths you have and plan what you need to do to help you achieve your goal.
  - The end of the year.
  - The end of next year.
  - The time you have finished school.
- Draw up a goal time-line to track your progress.

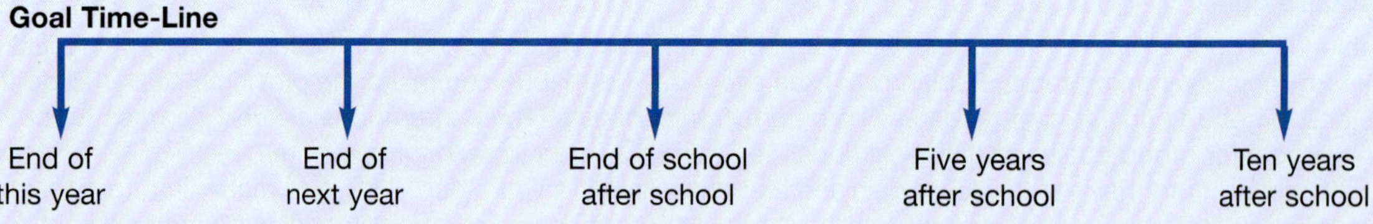

Knowing what you are good at also makes you feel good. It will make you want to engage in worthwhile activities. Think about what you are good at and how important this is to you, your family or the community. Perhaps you are good at looking after younger children or chopping firewood. This could be important to your family. Perhaps you are good at gardening, cash cropping, public relations, making canoes and fences. Know your strengths and take pride in doing these things well.

For you to try

- Think about what you are good at, and how important these things are to you. Make a list of things you are good at.
- Suggest how to use your strengths to benefit the community.
- Consider alternative ways to overcome your weaker abilities.
- A Undertake a grade survey to find out what students are good at. Show this information as a graph.

WAYS TO CONTRIBUTE

1. Time

Spending time helping others is important. Time is a resource – we can waste it or we can use it wisely. We can use it to benefit ourselves and we can use it to benefit others. Think about the past week. Think about time you spent wisely and time you wasted. Think about time you spent with yourself, with your family, with friends and with other groups in your community. Did it benefit you? Did it benefit others?

When we use time wisely, it benefits others and it benefits ourselves. Managing time is an important skill needed to achieve goals.

For you to try

- Make a personal timetable for a day or a week to show how you intend to use your time.
- Work out how much time you spent helping other children, family or community members. Did you use your time wisely?

2. Resources

Resources are anything we use. Natural resources can be found in nature. To use natural resources well, we need knowledge and skills. We also need technical resources. Tools and equipment are technical resources. Think about the resources you use each day to contribute to family and community life.

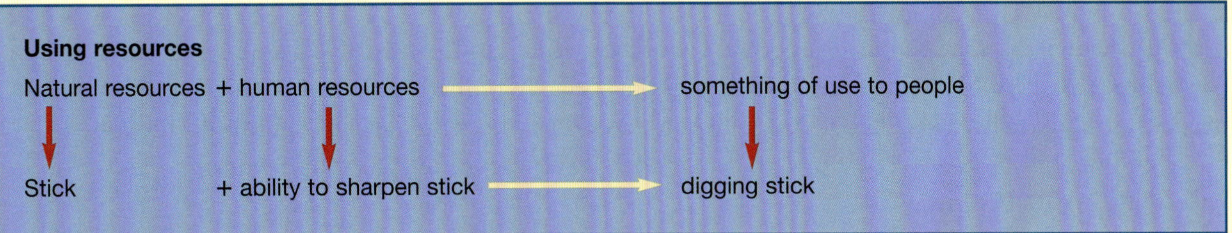

For you to try

- Go out into the nearby bush. Collect natural materials and make these into things that can be used by people.

- Collect used cartons, plastic containers, boxes, and so on, and make these into useful items.

3. Knowledge and skills

Attending school helps you to gain knowledge and skills that can be shared to benefit the community. You should look for opportunities to exchange ideas, to share, to show, and to help people think and work together.

You could help people make better food gardens using your knowledge of improved farming methods; help to prepare nutritious meals for their family; or teach people to mend torn clothes. You could even help your community to write letters or prepare a project proposal.

Helping in community work, using your knowledge and skills, will earn you respect and make you feel proud.

For you to try

- Take part in a community project by helping to carry out work, for example, putting stones in holes in the road.

4. Labour

Labour is the work that people do to help themselves, their families or the community. Labour can be skilled or unskilled. Skilled labour is when you are specially trained to do a particular job like a nurse, teacher or bank teller. Unskilled labour is when you do not need any special skill to do the job.

Everyone, old or young, male or female, can help their community through working hard to do the jobs that need to be done.

Teamwork in the community requires people to work together for a common purpose. We can work together to clear bushes, dig drains, collect items like stones or bush materials for building. We can work together to clean up a marketplace, beach or village. There is strength in people working together for the benefit of the whole community. When people work together, the job becomes easier.

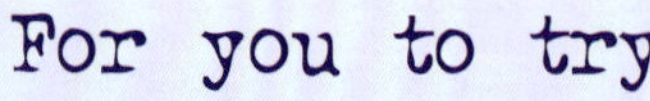

- Take part in a community village Clean Up Day, for example, coastal clean-up or market clean-up.
- Design and make a poster to promote community awareness to keep that local environment clean.

IMPORTANCE OF WORKING WITH THE COMMUNITY 

Community members use their skills and knowledge to help the development of our nation. We are all members of the Papua New Guinean community. It is important that we have the right attitude to work for the general benefit of all. It is important that we use our skills and knowledge to benefit the community and share our skills and knowledge with others. When we share, it shows that we care. It is important that we care about our community to bring benefits to others and ourselves.

We can contribute to our community through communication and interpersonal skills. Good communication can lead to improved relations with the community. But we also have to act. Actions speak louder than words. We have to do what we say we will do. We need actions as well as words to gain the trust of others and build good working relationships with them.

For you to try

- Identify someone you admire and respect in your local community.
- Identify reasons why you admire and respect that person.
- Invite that person to come and speak to your grade.

1. Promoting self-reliance

Self-reliance is about relying on yourself. It is about doing things for yourself, without always relying on other people. We need to use our resources wisely to meet our needs.

Traditional subsistence lifestyles required people to be self-reliant. It is important that we value our traditional ways and not lose them, especially the value of self-reliance. As we become a modern society, we cannot always be looking to the government or private enterprise to provide us with a living. We need to be able to support ourselves and our community.

For you to try

- Invite a community elder to tell how the community was self-reliant in the past.
- List the changes that have occured in your local area since Independence.

2. Improving living standards in the community

The quality of life should improve from one generation to the next. We should be leading healthier lives and living longer. Hopefully your life will be of a better quality than your parents and your children should have a quality of life that is better again. It is the responsibility of each one of us to help improve the standards of living in our community.

Consider how living standards have changed over time, in regard to water, lighting, food, clothes, education, transport, communication and travel. Have all the changes been for the better?

For you to try

- Invite a local level councillor to tell how living standards have changed from traditional time to modern time in the community.
- Identify and take part in an activity that will benefit your community, such as changing public places or cleaning common areas such as sport ovals.

Sepik River

Port Moresby city

POSSIBLE NETWORK PARTNERS

1. Government officers

It is important for communities to have good communication with government officers, such as fisheries, so that they can receive government services. Contact between individuals, groups and organisations is called networking. The *wantok* system has been a traditional network between people for a long time.

Network partners help one another in terms of:

- Sharing ideas and learning from each other.
- Sharing responsibilities and cooperating with each other.
- Listening to each other's problems and complaints and reaching an understanding together.
- Group discussions that lead to democratic decision making.

Information and ideas can be shared through face-to-face contacts, phone, letters, two-way radio or fax machines.

Communication with government officers is vital for communities to receive government services. Government officers are usually resourceful people who have had special training for their particular jobs. Experts from the local community can be invited to share their knowledge and experiences with others in our community.

Health workers like nurses, teachers, didiman/meri, police officers and district managers are some of the government workers who are network partners for community development.

For you to try

- Rice is a common food eaten in villages as well as towns. How could your school benefit from networking with a local officer from the Department of Primary Industries to trial rice growing? Explain your answer to the class.
- Find out the roles government officers play in your community and complete the table below.

Government officer	Roles played
1. Teacher	
2. Didiman/meri	
3. Health worker	
4. District administrator	

- Write a letter to invite a local forestry officer, based near your school, to talk to students on 'World Environment Day' celebrations.

2. Non-government groups

Non-government organisations (NGOs) are often set up with support from overseas donors. NGOs operate throughout Papua New Guinea. NGOs include missions and organisations such as Volunteer Services Overseas (VSO), World Wildlife Fund, Oxfam and World Vision. These groups work closely with government agencies and provide valuable skills and knowledge to assist community and national development. For example, many government and non-government organisations provide care for people with special needs.

For you to try

- Write to an NGO group working in your area to invite a representative to come and talk to students about the organisation's work and how the school can become involved.

3. Churches

Various church agencies have been in Papua New Guinea for over a hundred years and work in many different ways. They contribute much to the development of this country. They help with health care and the spiritual, educational, economic and social growth of individuals. Church leaders are often good role models for peace, social justice and how to work successfully in network partnership arrangements.

For you to try

- Invite a church pastor to tell the class about how the church supports community development.

4. Community elders and specialists

Networking between community elders and community specialists is important so that the skills, knowledge and wisdom of older people can influence decisions. Elders are a very special group of people in our communities. Through their involvement, respect is shown for the knowledge they have gained through long experience. They have seen many changes and developments in their lives. Changes that have been both good and sometimes bad.

For you to try

- Invite a community elder to speak to you about community changes over the years. Ask about the developments they would like to see happen. Discuss whether young people share the vision of older people and list how you could work together with local elders to develop your community.
- Take part in practical community projects organised by village elders or community leaders.

5. Local level government

Members of local level government are elected as ward councillors and district administrative officers. They form an important network of communication within the community. They share information on election programs, project funding assistance and community service work times. Local level government officers are key people in organising community projects.

For you to try

- Talk to your local council members and ask them to tell your class how community messages and government information gets passed on to the community.
- Role-play how village court magistrates deal with a case involving youths stealing from the village garden. Other roles to be played are: village court peace officer, elders and villagers.

6. Business community

Business groups and associations have important contributions to make to communities. Groups like the smallholder coffee growers' associations, poultry farmers' associations and public motor vehicle (PMV) operators can form important networks that contribute to community development. Communication occurs through meetings, radio programs and awareness campaigns in public meeting places.

For you to try

- Identify businesses that operate in your community. Discuss how networking between business and the public assists development.
- Hold a grade discussion to consider how information (such as current prices of produce, trading hours, and other important information) gets passed to members of the community?

ESTABLISHING CONTACT WITH NETWORKS

1. Identifying network partners

Before your school can establish contact with networking partners, you need to identify who they are. Find out the names and contact details of networking partners in your community:

- Government agents
- Non-government groups
- Churches
- Community elders
- Community specialists
- Local level government members
- Business people.

There are many benefits for the community in establishing network partnerships if everyone in the relationship is honest and is willing to accept new ideas.

For you to try

- Find out and list contact details for groups in the community with whom the school could network to undertake community development projects.
- List projects that could be undertaken jointly with different community groups.

2. Make initial contact with partner organisations

Networking is based on human contact between individuals and groups. Forming relationships through gatherings provides chances for contacts to be made with individuals who represent the different groups and organisations. Meetings, workshops, short courses, social clubs, church fellowship nights, sports meetings, and school P&C meetings all bring people together.

Other contacts can be made through letter writing, sending fax messages and e-mail, talking to someone on the telephone or meeting people face to face. When you make your first contact, proper interpersonal skills are needed. When talking to someone that you do not know for the first time, on the phone or in person:

- Introduce yourself briefly.
- Speak clearly and look at the person you are talking to.
- Do not interrupt when the other person speaks.
- Ask questions if there is anything you would like to know.
- Thank them for their time and help.
- Promise that you will contact them again when the need arises.

For you to try

- Role play the skill of meeting new people. This can be done in pairs or small groups (3–4 members).

3. Organise meetings

Community meetings bring together resources, ideas and support for community development projects. Interested individuals and group representatives can all contribute to discussion at meetings.

Meetings are important gatherings where people discuss matters affecting them and the community. Meetings help to pass on advice and information. Action plans can be agreed to carrying out the recommendations of the meeting. Meetings should allow everyone to take part.

For you to try

- Role play: conduct a meeting to improve the classroom environment. One student should be the chairperson and one take minutes. Everyone should be given opportunities to express his or her view.
- Read out each group's decision and action plans for other students to hear and make comments.

4. Use telephones

If you need to pass on or find information quickly, you can use the telephone. To use the telephone well, you should practise speaking clearly.

When making telephone calls you should:

- Dial your number and listen for the dial tone.
- Wait till the receiver is picked up.
- Greet whoever answers and ask to speak to the person you would like to contact.
- When the person you want is on line, communicate your message to him/her.
- Listen carefully too and make appropriate responses without disruptions.
- After all is said, thank the person and end the conversation.
- Gently replace the receiver on the phone.

Keep an accurate record of telephone contact numbers of those people and organisations that you need to ring often, then it is easy to call your network partners.

For you to try

- Role play: practise making telephone conversations using strings with two tins attached.
- Carry out a quick survey in your class to find out the number of students that have used telephones. Ask them to share their experiences.

5. Write letters

Letter writing is one common way of communicating with network partners. Write letters to your network partners or contacts when:

- You cannot contact them in person or by telephone.
- You want to make communication official and confirm earlier personal or telephone conversations.
- You require extra information.

Writing business letters is different from personal letters. When writing a business letter, remember to make it short and clear. Personal letters are written to friends and families; they can be long and do not follow a set format.

For you to try

- Write a formal letter on school letterhead paper, inviting an important network partner to visit your school. Ask him or her to give a speech on how to network with organisations.

Community Projects

PLANNING STRATEGIES AND PROCEDURES

When starting out on any project the following steps should be taken.

1. ***Investigate***. Ask questions to find out what are the specific needs of the school community and how these can be met. This information can be collected by:

   - Writing a questionnaire and seeking opinions from students.
   - Conducting a survey.
   - Listing five possible projects and asking students to vote to decide which project is top priority.

2. ***Plan and design.*** Once you have decided on the project, you need an action plan. Students need to be organised into groups and each group should be given a specific job. At this stage students must be able to:

   - Design the object (e.g. fence for school garden) or school improvement project (e.g. repairing school desks). Write a list of resources needed to make or complete the object.
   - Write the steps for completing the project as well as a proposed time-line for completion.

3. ***Implement the plan***. Work cooperatively with the team to complete the task. Appoint a team leader who will:

   - Give every member of the team a task.
   - Make sure the project is progressing to plan.
   - Manage time and resources effectively.
   - Address problems if they should occur.

4. ***Evaluate.*** Once the project has been completed, it is important to evaluate the project.

   - How clear were the plans and procedures?
   - Did the students work as a team?
   - How reasonable was the schedule?
   - What were some of the problems that made it difficult to complete the project?
   - How will students use this school improvement?
   - Can this project be extended or developed further?
   - Write a report on the processes involved in the project.
   - If money was involved, write a financial statement for the project.

1. Cooperative projects in your community

Haus win (house wind)

A *haus win* is an open hut or building built apart from other building as a relaxation place. Most rural homes are completely enclosed for security reasons or cultural value. This limits fresh air and sunlight. A *haus win* in your backyard garden, or beside your home, is an ideal place for entertaining friends and visitors. Most families in coastal villages have a *haus win*. The style and type of *haus win* varies from region to region in Papua New Guinea.

For you to try

- Sketch a drawing of the type of *haus win* you would like.

Public noticeboard

A public noticeboard is a community information place where community notices, *tok saves* and news are placed for the public to read. Public noticeboards are usually found where people frequently visit, in front of churches, post offices, stores, and community halls. In schools and large communities, a public notice-board is a very useful place for people to go to read information.

CENTRAL TB CLINIC SCHEDULE

MONDAY - 8AM-11AM - MEDICINE SUPPLY	1:30PM - 3:00PM - WARD PATIENT DISCHARGES
TUESDAY - WARD ROUND	ADULT TB PATIENTS REVIEW
WEDNESDAY - 8AM-11AM MEDICINE SUPPLY	TB DEFAULTER COUNSELLING
THURSDAY TB PATIENT UNDER 12YRS REVIEW	ADULT TB PATIENTS REVIEW
FRIDAY - medicine supply - FAMILY SCREENING FOR ADULTS	WARD PATIENT DISCHARGES CLEANATION -

NB. OUR VOLUNTEER PACK TB MEDICINE FOR TB PATIENTS

THANK YOU - SK TBC.

CLINIC VISITS

MONDAY MORESBY SOUTH — KILAKILA CLINIC ✓, VABUKORI ✓, BADILI ✓, LAWS ROAD ✓

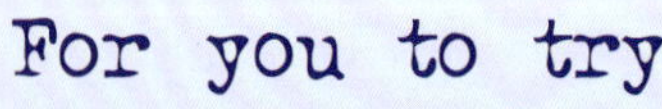

For you to try

- Visit noticeboards in your towns and make a list of information placed on these noticeboards.

- Design a simple noticeboard and build it as a class project.

Plant nursery

A plant nursery (also called a greenhouse) is where seeds and young plants are placed in pots or plastic bags to be cared for until the plants gain strength to be transplanted. The roof and the walls of the greenhouse are covered with a canopy screen to prevent strong sunlight and hard raindrops damaging young plants. People also place plants in pots and keep them in nursery houses and then sell them for money. Other plants in pots can be taken into homes and offices to make them look attractive.

For you to try

- Go for an excursion and visit plant nurseries. Study the types of plants and types of pot plants.
- Build a nursery house for yourself, or for your class or school.
- Collect pot plants, seedling bags and preferred plants for your nursery house.
- Sell some of your pot plants and take some to your classroom or your home.

Road maintenance

Roads are the link between homes, villages, schools, towns, villages and many other places. There are different kinds of roads, but all roads must be maintained. Maintenance of important roads is the responsibility of the provincial and national governments. However, the local and national governments do not maintain most of the smaller roads that lead to our communities, schools and towns. When roads are in bad condition, it is difficult for vehicles to travel. Bad roads make it difficult for rural village people to transport food to markets and bring back other goods and materials to the villages. People in local communities should be responsible and maintain these smaller roads.

For you to try

- Go on an excursion to inspect the road leading from your village to your school, or leading to the main road to town.
- Take note of any section of the road that is damaged and note what caused the damage.
- Report your finding to your teacher, head teacher, or the community councillor and village leaders.
- Plan ways to involve the whole school or the community to carry out maintenance work.

Water supply

Water is one of the essentials in life. Clean and healthy water to drink, wash and cook with is very important. The type of water supply system in communities varies from place to place. Urban communities use town water supply. Permanent houses may have their own tanks installed to collect water. However, most rural villages in Papua New Guinea still use creeks, rivers, springs and water wells to obtain water. Some communities are lucky to have a water supply system installed in their communities. Water pumps, tanks, dams and wells can be installed or built by government and non-government organisations.

For you to try

- Describe your school's water supply system. Propose a plan to improve this system. Draw a diagram of your proposed system.

2. Community-driven initiatives

The expression 'no man is an island' suggests that no one can live all alone. Whether you come from a village, town or city, you belong to a community. The first community is your family. But there are many other communities to which we belong, and each of these has different needs, roles and responsibilities, for members. We may belong to church groups, sporting groups and school communities. To be useful members, we take part in decision making or work on community development projects to promote self-reliance, self-esteem and ownership.

For you to try

- Make a list of the different communities that you belong to. Write about the roles you take in these communities.
- List needed projects for your school or local community and develop an action plan.

3. Work with a partner organisation

Partner organisations have been set up in some communities. These partner organisations are there to give advice and work with the local community on specific projects. Examples of these partner organisations are non-government organisations (NGO), church organisations, donor agencies (often sponsored by donor countries), business firms and governments.

For you to try

- Name some of the NGO groups in Papua New Guinea. What are their roles?

- Write a letter to any one of these organisations. Ask for booklets, posters, and other information about their organisation. Invite one of their representatives to come to your school to give a speech.

SHOWING INITIATIVE

1. Approaching community groups with common interests

Many community groups have been set up by the government, churches or non-government organisations to help people in communities solve problems and to provide for their needs. These groups or organisations help establish community projects, and carry out awareness campaigns. Many of these organisations are non-profit making groups. Their main aims are to improve the way we live, to encourage healthy living and protect destruction of the natural environment.

For you to try

- Conduct a survey and identify various community groups in your local area. Ask about the roles of these groups.
- Invite community group leaders to your school or class to explain their work within the community.
- If there are specific projects running at your school, invite these people to give advice. Write a letter to the organisation asking them to help your school complete the project successfully.

2. Involving community members in generating ideas

It is important to involve community members in helping to improve the lives of the community. You can do this by watching and listening to people, asking questions and gathering information, conducting a survey and identifying people's problems and needs. This gives people the opportunity to have a say or express their opinion on changes to and development of the community.

For you to try

- Organise public meetings or forums on current issues or development activities in the community and around the country.

3. Organising ideas and working out ways to achieve common goals

Once information is gathered and ideas are generated about how to improve community life, plans and proposals must be prepared. People must agree on what is to be done and in what order. People have to make detailed plans about how particular projects should be carried out. They have to decide which resources to use, and how to organise the labour, money, tools and other materials to successfully complete the project.

Class Project

- Plan an income-generating project for your school.
- Work out the cost involved, materials and tools needed and the labour or people required to run the project.
- Carry out the project.
- Is the project successful? Is it generating a profit?
- Record progress and aim to improve in the areas you identify as being weak or a failure.

4. Working together to develop plans

It is always wiser to work as a team. Involve community groups such as women's groups, youth groups or local sporting groups. Seek assistance from partner organisations such as the NGOs, church groups, government organisations or business firms. These groups can assist in money, materials, skills and expertise to implement community development. Every year there are funds made available through these organisations for community development.

Plan a cooperative project

Large-scale development projects involve many community groups and organisations. It is important to develop an action plan to involve all communities, clans, schools, religious and cultural groups. Such involvement strengthens cooperation and unity among various groups. The benefits are equally shared.

For you to try

- Identify needed projects in your school or community.
- Discuss with your teachers, head teacher, school board of management group, councillor and other community leaders.
- Approach partner organisations for funding and assistance.

4 GRADE 8 Managing Resources

Chapter Summary

In this chapter you will have an opportunity to:

- ✔ Investigate and evaluate current land and water resource management in your local area, and plan a small project identifying appropriate management practices that will generate income.
- ✔ Investigate and implement environmentally friendly ways of managing your local environment.
- ✔ Plan a crop or animal project for your local area that will be aimed at generating an income.

Syllabus References

Strand: Managing Resources

Substrand 1: Land and Water Management

Outcome 8.1.1: Evaluate current practices of land and water resource management to design sustainable resource management projects to generate income.

Substrand 2: Environment

Outcome 8.1.2: Describe and reflect on the economic, cultural and ecological values of natural, social and built resources and apply environmentally friendly ways of managing the environment.

Substrand 3: Crops and Animal Management

Outcome 8.1.3: Plan, design and implement a crop or animal management project suited to local conditions and using local resources, and aimed at generating an income.

Land and Water

CURRENT PRACTICES

In Papua New Guinea, more people depend on agriculture than on any other type of work.

1. Subsistence farming: gardening and fishing SS

Farming/gardening

Subsistence agriculture means growing enough food for you and your family to eat. Today, there are only a very few people in the remotest areas of Papua New Guinea who are sole subsistence farmers. The introduction of money, or the cash economy, has brought about rapid change.

It is commonly accepted that subsistence farming is growing food and materials to support the life of the farmer and his or her immediate family. All the food that is grown is usually used up. Any food left over can be sold to buy essential items that the family needs.

There are many different types of subsistence farming in Papua New Guinea because of the many different environments, customs and values.

Shifting cultivation is the main system by which food is grown throughout Papua New Guinea. It can take two forms:

- Clearings from forest.
- Clearings from grasslands.

The main problem with shifting cultivation is that if the land's resting time is too short, the soil does not recover its fertility. When this happens, food yield per unit of land is low. To increase crop yield, people need to find different ways to keep the soil fertile.

Subsistence farming involves:

- Crop rotation.
- Planning new garden sites.
- Use of traditional tools.
- Different roles for those involved in gardening and fishing practices.

The combination of these efforts ensures a successful harvest.

Today, subsistence-farming practices are changing because there are more and more people living in the country. New farming methods are being introduced and these new methods, when combined with the old ways of farming, benefit everybody. This is very important because the country's population is growing and farmers must grow more crops and raise more livestock to feed their own families and also sell to markets to feed urban dwellers.

Highland areas: Highlanders have developed an intensive farming system. They use land for long periods of continuous growing and, where necessary, use fallow land after as little as two years. They enrich the soil with decaying matter and rotate crops to help keep the soil fertile. Most highlanders plant on mounds. Mounds concentrate nutrients, drain off water, and help protect the plants from frost.

Foods planted by subsistence farmers in the highlands include: sweet potato, taro, sugar cane, pitpit, winged beans and many Chinese and European vegetables.

Lowland areas: A system of shifting agriculture and bush fallow is used in most lowland areas. Garden lands are cleared and the bush is burnt to return nutrients to the soil before crops are planted. Taro harvesting can begin after three months, yams after six or seven. The land may be used for six months to three years. Then it is abandoned and natural bush is allowed to grow. This bush fallow restores nutrients to the soil. The length of the bush fallow depends on the soil's fertility and also on how much other land is available. It ranges from seven to forty years, usually more than fifteen years.

Foods planted by subsistence farmers in the lowlands include: sago, taro, yam, banana, cassava, coconut, breadfruit, pawpaw, mango, marita, galip, okari nuts and greens.

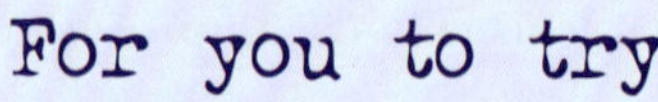

- In small groups, describe the different roles played by men and women when subsistence farming, gardening and fishing.
- Invite a speaker to talk about how people in your community harvest huge yams, sweet potatoes and taro.
- Draw and describe some traditional gardening and fishing tools.
- Ask parents and elders about traditional songs sung during the yam, sweet potato and taro harvests. Learn these songs, and any other harvest rituals.

Traditional fishing

In Papua New Guinea, most people have traditional fishing grounds where they go to catch fish. The fishing grounds may be close to a lake, river or next to a reef. In coastal areas, local fishermen go out at night to fish using pressure lamps and spears, an underwater torch and fishing guns, or a string line and hooks. Most of the fish caught are either smoked or taken fresh to be sold at the market for cash. Fishermen purchase essential goods from trade stores using the money earned from selling fish. The Koki market in Port Moresby is one place where local fishermen sell their fish for cash.

As the population grows, more and more people are using illegal and damaging fishing techniques to increase their fish harvest.

- Some coastal fishermen use dynamite to catch fish. This destroys fish breeding grounds.
- Inland fishermen sometimes use poisonous derrie roots to catch freshwater fish, prawns and eels. The poison in the water also destroys hundreds of young fingerlings and prawns.
- People fishing in lakes and rivers sometimes use gill nets and these reduce the fish population.

For you to try

- Invite a village elder to explain traditional fishing methods in your local community. How environmentally friendly are these fishing practices? Discuss.
- Propose an action plan to the Fisheries Authority that will stop people from using dynamite to catch fish on coral reefs.

2. Improving subsistence farming and gardening SS

Subsistence-style gardening is the foundation of today's agriculture. Our task is to integrate these gardening practices into workable agricultural programs for sustainable development.

It is important that subsistence agriculture in Papua New Guinea continues to be more efficient so that more food can be produced to feed a growing population. In addition, subsistence-style farming should develop alongside commercial farming to grow cash crops for export, which will earn money for the Government.

People should be encouraged to:

- Use better farming tools to aerate the soil. This ensures that soil nutrients are well mixed and also slows the growth of weeds.
- Dig out weeds and mulch as this adds nutrients to the soil.
- Use a fertiliser or manure on gardens to increase crop yield.
- Feed animals the appropriate foods to ensure growth. All animals, and freshwater fish, will flourish if fed the 'right' food.
- Learn how to manage livestock so that stock levels increase each year.
- Maintain roads so that crops and livestock can get to markets.
- Make farming resources, such as tools, fertilisers and seeds, readily available at a reasonable price so that people can purchase and use them.

For you to try

- Observe agriculture in your area. Ask questions of people in your community and write a report.
- Discuss the importance of improved subsistence agriculture for your province.
- What improvements could farmers in your province make to raise their standard of living?
- A Identify ways of producing a high crop yield. Design a poster to highlight these ways.

SS 3. Commercial farming and fishing

Some farmers in Papua New Guinea only grow cash crops. They plant a crop that they won't use, but someone else wants to buy. Cash crops (crops that are planted specifically to raise money) are grown by different groups of people in different ways.

- **Plantations:** These are large areas of land, usually owned by foreigners or companies. On plantations, machinery is used, and labour is often brought in from outside the local area. In Papua New Guinea, plantations grow copra, cocoa, coffee, rubber and tea.
- **Smallholders:** These are usually subsistence farmers who also grow cash crops to raise money. In the Highlands the main cash crop is coffee with some pyrethrum and cardamom. In the lowlands cocoa, copra, some coffee and rubber are grown.
- **Nucleus estates with smallholder settlers:** These are schemes whereby a large estate is owned by a company and surrounded by lots of smallholdings that grow the same crop as the estate. The smallholders can use the company's machinery and sell their crops to the company.

Most rural people are both subsistence and cash crop farmers.

In addition to agriculture, people also raise fish, pigs, chickens, cattle, goats or sheep, which is another kind of modern, cash-farming enterprise.

Most of the produce from cash crops is exported to overseas markets. Growing cash crops requires a large number of labourers or machinery as well as large areas of land. In addition, higher management and accounting skills are required for the smooth operation of the farm. The overall aim is to make huge profits to expand the farm or finance new developments. Although cash crops bring in much needed money for our country, some aspects of commercial agriculture create problems that need to be addressed.

- **Soil erosion:** Soil and its nutrients are essential for plants to grow. As people clear forests to grow cash crops, soil erosion occurs because there are not enough plant roots and organic materials in the soil to keep it in place during wind and rain. Once the topsoil, which contains all the nutrients, has been washed or blown away, the soil becomes infertile and crop yield is very low. Many plantations are addressing this problem, but much more needs to be done to ensure forest clearing is regulated. Reforestation programs need to be put in place.
- **Destruction of traditional hunting grounds:** Traditional hunting is still a way of living for many people, and the destruction of these grounds means that people have lost one form of food supply, as well as materials for everyday living and ceremonial artefacts. Government regulations need to balance economic development with traditional ways of living.
- **Water and air pollution:** All large-scale agricultural enterprises bring with them environmental concerns, such as air and water pollution. Water pollution from chemicals used in farming is a common complaint and air pollution, especially from large-scale mining, is evident in some regions. Regulations need to be set in place and enforced.

Fishing is a small industry in Papua New Guinea, although our country has excellent fishing waters. Fisheries estimate that Papua New Guinea's seas could produce 500 000 tonnes of fish a year without risking future stocks. There is also a sizeable potential market, but despite this, the fishing industry has not grown significantly. Why?

- Fish prices are low compared with the costs and risks involved in catching them.
- Traditional land rights sometimes prevent fishermen from entering the best fishing or bait-gathering areas.
- Harvesting prawns, lobster tails and barramundi, and tuna fishing, are mostly in the hands of foreign companies. Their boats have specialist equipment for large-scale fishing.

For you to try

- Invite a government officer to talk about commercial agriculture and fishing. Write an information report.
- Identify and discuss examples of poor agricultural practice by foreign commercial companies.
-  Write a pamphlet to explain good commercial fishing or agricultural practices.

4. Terrace farming

Terrace farming is new to Papua New Guinea. In South-East Asia, the large populations in mountainous areas have forced people to invent new ways of clearing enough land to grow crops. Where the land is too steep for ordinary cultivation, terraces have been built. Natural woodland has been cut down and large steps have been cut into the steep slopes. Small walls, built along the edges of the terraces, prevent soil from being washed away. Developing a terrace system is hard work, but it enables farmers to plant crops on hillsides and it makes crop harvesting and transportation to market easier. Terrace farming also prevents and controls soil erosion and landslides.

In Papua New Guinea, and other Pacific islands, people leave the trunks of fallen trees lying across steep slopes. The trunks are held in place by stakes driven into the ground. Although not really terraces, this prevents soil from being washed away.

For you to try

- If your school is located in the hills, plan a terrace gardening system around a steep hill.
- Use the library to read books about terraces in other countries and identify the types of crops best grown using this system.

5. Traditional fishing, hunting and gathering practices

In Papua New Guinea animals are hunted for useful reasons and not for sport. Hunting is a minor source of food when compared with the numbers of people gardening, gathering, fishing or raising pigs. Hunting does feed many people. However, hunters provide many useful materials for village life. Feathers and fur are used for decorations. Skins are used for drumheads, or, in the case of crocodiles, for money. Bones are used for spear points, scrapers and other tools.

Men and boys do most of the hunting using weapons. Men hunt the larger animals – crocodiles, wild pigs, cassowaries, wallabies, tree kangaroos and dugongs. Women and children catch small mammals such as lizards, frogs, spiders and insects. Snakes are hunted by some people, but avoided by others.

Traditional hunting practices include:

- Using traditional hunting weapons for traditional purposes.
- Ensuring that female animals are not caught so that they can breed and save the species from extinction.
- Clearing bush and cutting down trees only if absolutely necessary.
- Using clan landmarks or boundaries for hunting to avoid tribal disputes.

Traditional fishing is an important job in Papua New Guinea. One-fourth of rural families fish. Papua New Guineans consume more than 15 000 tonnes of fresh fish each year. Fish are the main source of protein for most coastal and river people. Both men and women fish.

Some good traditional fishing practices include using:

- Bows and arrows to shoot fish from rocks or canoes.
- Dip and scoop nets with frames of bamboo or wood. These are passed through the water, along the shore or from a canoe.
- Hooks and lines. In many places, metal hooks have replaced the traditional hooks made of turtle shell, bone, wood or other material.
- Fishing nets worked in several ways.
- Plunge baskets, best in muddy water.
- The traditional calendar is followed to hunt certain species of marine life for personal use.
- Fishing practices, such as the use of explosives, are used under strict conditions as coral reefs can be easily destroyed.
- Traditional taboos that limit access to fishing grounds are respected and this makes sure that fish numbers grow.

For you to try

- Discuss, in groups, the traditional practices of fishing and hunting in your community. What are the set rules for fishing and hunting?
- Invite an elder from the village to explain and make some traditional hunting, fishing or agricultural tools.
- Gather resources and make one of these tools. Where possible, go into the bush or water to try using these tools.

SUSTAINABLE RESOURCE MANAGEMENT

1. Permaculture SS

Permaculture is 'permanent agriculture' or sustainable agriculture. Plant growth is kept up or continued by plants regenerating themselves repeatedly.

Permaculture emphasises:

- Perennial plants rather than annual planting of crops. This reduces soil erosion as there is no ploughing the soil.
- A mix of plant sizes and types planted together. Often one plant provides good conditions for the growth of another plant species. Weeds and pests are less of a problem (see below).
- Plants and animals are farmed together. Animals can feed on grasses and excess or waste plant produce. In turn, animal manure can be used to fertilise plants.
- Local production for local consumption reduces the need for transport and the pollution that comes from transport. Permaculture takes care of the environment and encourages self-sufficiency and sustainable living.

For you to try

- Invite an agricultural officer to talk to students about permaculture and how it helps the livelihood of the people in your community.

2. Mixed cropping

Mixed cropping is a way of planting various crops in the same garden at the same time without any definite spacing and without rows or lines. Mixed cropping is the traditional way of planting crops in Papua and New Guinea and most parts of the Pacific region. It is an easy method of planting and it provides a variety of crops for harvest all year round.

Each mixed crop must contain at least one of each of the following crop categories: leafy legumes, tuberous and fruit-bearing vegetables. In this way, the nutritional needs of a family are met by growing a diversity or mix of vegetables that mature at different times. In this way, a family has vegetables available throughout the year. This practice also helps with pest control as certain intercrops act as insect repellents. One disadvantage is that crops tend to be too crowded and this can affect crop yields.

For you to try

- Invite an agricultural officer to your class to talk about the advantages and disadvantages of mixed cropping.
- Visit local gardens to see a mixed cropping system.
-  Plan a school garden using the mixed cropping system. What will you plant to ensure that the garden will produce food that provides all your nutritional needs?

3. Organic farming

This is an old concept that is used in many primary schools. Organic farming means growing crops using compost, manure and other natural plant foods, without the use of chemical fertilisers. This is the cheapest and best method of growing crops for human consumption. Schools are encouraged to farm organically because chemical fertilisers can be very expensive and dangerous to use.

For you to try

- Practise organic farming in your garden plots.
- Invite an agricultural officer to talk about the advantages and disadvantages of using organic farming methods.
- Make comparisons between organic farming and the modern use of chemical fertilisers.

Class Project

Objective: Plan a school garden using the mixed cropping system.
When starting out on any project, the following steps should be taken.

1. ***Investigate.***
- Ask questions to find out about soil composition and preparation.
- Determine the best location for your garden. Factors to consider include: a suitable flat site, water supply, protection from animals and type of farming.
- Ask your local community about the best plants to be included in your garden.

2. ***Plan and design.***
- Once the project has been determined, there is need for a plan of action. Students need to be organised into groups and each given a specific objective. At this stage students must be able to:
    - Write steps for completing the project as well as a proposed time-line for completion. This will include garden preparation, planting procedures, garden maintenance, pest control.

3. ***Implement the plan.***
- Work cooperatively with the team to complete the task. Appoint a team leader who will:
    - Give every member of the team a task.
    - Make sure that the project is progressing to plan.
    - Manage time and resources effectively.
    - Address problems if they should occur.

continues

4. *Evaluate.*

Once the project has been completed it is important to evaluate the project.

- How clear were the plans and procedures?
- Did you work as a team?
- How reasonable was the schedule?
- Were there any problems that made it difficult to complete the project?
- Will this plan help you develop your own mixed-crop gardens?
- Can this project be extended or developed further? Based on this project, do you have any suggestions?
- Write a report on the processes involved in the project.

4. Reusing and conserving resources SS

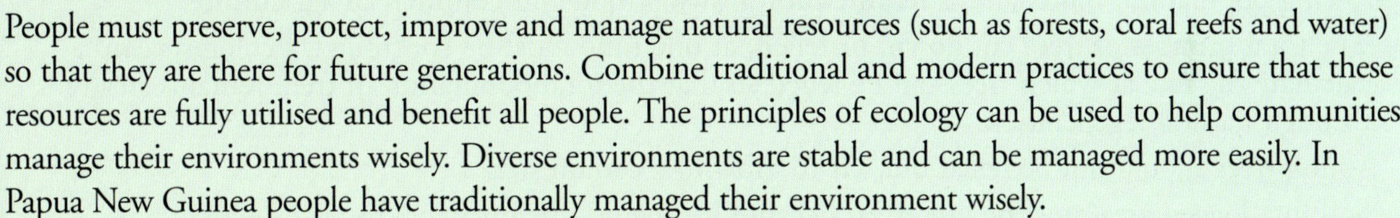

People must preserve, protect, improve and manage natural resources (such as forests, coral reefs and water) so that they are there for future generations. Combine traditional and modern practices to ensure that these resources are fully utilised and benefit all people. The principles of ecology can be used to help communities manage their environments wisely. Diverse environments are stable and can be managed more easily. In Papua New Guinea people have traditionally managed their environment wisely.

Examples of wise management practice

1. The coastal villages in Manus have community laws to protect areas where green turtles and dugong (sea cow) live and breed. This is because these species of marine life are very rare. These marine protected areas remain untouched so animals can maintain their numbers.

   When people use lagoons and coral reefs in a sensible way, the reefs renew themselves naturally, the fish population is maintained and we have what is called 'sustainable' development. In this way, only selected species of fish are harvested, smoked and sold at the Lorengau market for cash.

2. The government has established National Parks to preserve native animals and plants that are significant to Papua New Guinea. Due to increased fishing, hunting, gardening, logging and mining activities, some plants and animals are in danger of extinction. The government has declared these 'National Animals' in order to protect them. They can only be caught or collected by Papua New Guineans using traditional methods and they can only be used for traditional purposes.

3. The government has also established wildlife management areas. In these areas local landowners establish their own rules in order to limit damage and protect the natural environment. These rules are usually based on traditional management. If the environment is not used and managed wisely, more and more plants and animals will become endangered.

For you to try

- A Design a poster to promote a sustainable rainforest environment or marine life protection.
- A Write a report outlining the best ways of protecting the local natural environment.
- Invite an environment and conservation officer to your school to talk about the different ways native animals and plants are being protected.

Environment

ECOLOGICAL VALUES

1. Forests

Forests are important because they support and protect the soil. They are home to many plants and animals that are part of the forest ecosystem. Forests conserve water and provide river catchments. They are a place for leisure activities and the source of many products used by people.

Good virgin forests take many thousands of years to develop, but they can be easily wiped out in a year.

Forests are now being cut down faster than they are regrowing or regenerating. This is because, with a growing population, more land is required for growing food and commercial agriculture.

- The total number of trees cut down, or felled, at one time must be controlled. Ecologists can calculate the rate at which trees can be felled so that the total population of trees remains steady through natural regeneration.
- Young trees should not be felled but left to grow to a certain size.
- Soil in a logged area must be preserved because it provides the nutrients for re-growth.
- Care must be taken to ensure that steep slopes are not logged because of the risk of soil erosion.
- Strict control over commercial logging companies is required. (New Ireland has already lost almost all of its forest resources because of poor management.)

For you to try

- Invite a forestry officer to your school to talk about forest management practices in your community. Write a report.
- Interview people in the community about how they look after their forests.
- Plant trees around your school on Environment Day.

2. Coral reefs

The seas and coral reefs are an important resource. They contain enormous numbers of sea creatures such as prawns, coral and reef fish. If too many of these sea creatures are caught, a valuable source of food and income could be lost forever.

Traditional methods of catching fish did not remove large numbers of fish at one time. However, large commercial fishing boats are now catching huge quantities of fish using modern fishing equipment. Management must keep strict control over these companies to safeguard the fisheries for the future.

- Regulations need to be introduced to prevent over-fishing.
- The total number of fish caught needs to be limited. (Ecologists can calculate the maximum number of fish that can be caught without risking a decline in the population of that species.)

- The number of commercial fishing licences must be limited.
- Young fish should not be caught as they are breeding stock.
- Fishing could be restricted at certain times of the year, when fish are breeding or migrating.
- Trawling should be banned.

For you to try

- Visit a big river or reef for an excursion.
- List positive ways to care for our coral reefs.

3. Importance of an ecological system

Ecology is the study of plants and animals in relation to their surroundings. Many living things interact because of their need to feed.

1. Plants absorb nutrients from their non-living surroundings, which they use to make food by the process of photosynthesis. They are known as producers.
2. Plants may then in turn be eaten by herbivores (animals who depend on plants for their food).
3. Herbivores can be eaten by carnivores (animals who eat other animals for food). These are known as consumers.

This feeding relationship between plants and animals is known as a food web.

Living things also depend on their non-living surroundings. Plants need water, energy from sunlight, carbon dioxide from the air, nutrients form the soil, and shelter from their surroundings. Animals also need water, sunlight, oxygen from the air, and shelter from their surroundings.

The relationship between living things and their surroundings is often complicated and people who study these relationships are called **ecologists**. In order to survive and not destroy the environment that supplies our needs, we must understand the ecological system.

For you to try

- Go into the bush and observe the ecological system. Draw diagrams and food chains (like those on page 4) to show the interaction between all the living things (plants, insects, animals) you observed.

ECONOMIC VALUES

We change our environment by farming the land, harvesting the forests, fishing the seas and mining the earth. We do this to satisfy our basic needs and to improve our quality of life. However, as our population increases these needs also increase. If our environment is to continue providing resources for people, and not be destroyed, we must look after or manage it well.

We must use sensible management policies and procedures that apply the principles of ecology to maintain a stable environment. For example, if prawns are being harvested, management should make sure that the total population of prawns remains stable. If too many prawns are harvested, the population will decline and eventually the species will die out.

For you to try

- Invite a guest speaker from the tourism industry to talk about tourism opportunities.
- Make a list of the kinds of plant life and living animals around your school and immediate location.
-  Discuss possible ways of earning money from things in your environment, for example, natural materials for making artifacts, wood for making toy canoes.

CULTURAL VALUES 

The natural environment is important for cultural reasons. Many people use materials from forests and seas to make weapons for use in tribal customs and ceremonial rituals. Hunters sometimes capture animals alive for ritual purposes, as well as nutrition. As the forests are cleared and products are developed for export, many traditional materials are becoming scarce. The government must regulate export quotas of specific materials, including plants and animals, to ensure the survival of Papua New Guinean culture.

For you to try

-  Go out into your local community and make list of plants and animals that are used for cultural ceremonial activities.
- Arrange with village elders to visit the sacred places in your community.
- Take part in traditional singsings or ceremonies.

Crop and Animal Management

VALUE OF CROPS AND ANIMALS

Papua New Guinea is very rich in natural resources. Many of these resources have an economic, nutritional and cultural value and they are highly valued by communities.

Most crops grown in Papua New Guinea are not native to this country. Coconuts were introduced by Queen Emma (an early pioneer in east New Britain), and early European settlers introduced several crops including sweet potato, yam, cabbage, cocoa and taro. The most important crops, believed to have originated from Papua New Guinea, are the winged bean and sugar cane. Two important native tree crops are the Hoop pine and Klinki pine.

PD 1. Nutritional value

Significant crops grown in Papua New Guinea can be categorised as follows:

- **Tubers:** a root crop that is our most important source of carbohydrate. The main tubers grown are sweet potato, yams, taro, potato and tapioca.
- **Fruits and nuts:** our main source of vitamins and fat for the body. The main fruits include banana, pineapple, pawpaw, passionfruit, citrus, guava, avocado, mango and breadfruit.
- **Vegetables:** our main source of vitamins for the body. The most common vegetables include cabbage, eggplant, lettuce, tomato, onion, cucumber, cauliflower, broccoli, oenanthe, pumpkin and silver beet.
- **Grain legumes:** a source of energy for the body. The most common grain legumes include peanuts, various beans (including the winged bean), cowpea, soy beans and peas.
- **Spices:** used mainly for seasoning. These include cardamom, ginger, turmeric, peppers and chillies.
- **Tree crops**: these are called permanent crops, cash crops or perennial crops. These are crops that grow and produce for many years and they include coffee, coconut, cocoa, rubber, oil palm and citrus.

For you to try

- Go into your local community and observe and record the crops and animals that you see.
- Look at the table below and, based on your own knowledge, fill in blank spaces.
- Add other crops and animals that you observed to this table.

Important crops and animals

	Crops	How it is used	Animals	How it is used
1	Yam	A high protein food, included in many meals	Pig	
2	Banana	A high energy food; a common family food	Cattle	
3	Cocoa		Chicken	
4	Coffee		Sheep	
5	Coconut		Fish	
6	Vanilla		Cassowary	
7	Taro		Turtle	
8	Winged bean		Goat	

2. Customs and beliefs PD

People are a country's most important resource. Our skills, knowledge and respect for the land on which we live have enabled us to live and develop it over many thousands of years. These things are part of our traditions.

- **Agriculture:** Subsistence farming and gardening is mainly carried out by women and children. Men are involved in clearing the land. Girls are taught to maintain gardens and cook. Much of the social, festive and daily living activities revolve around subsistence agriculture. The traditional village calendar highlights this. Festivals are common at times of planting and harvesting crops. These festivals involve dancing and singing, the beating of drums and blowing of conch shells. Musical instruments are made from wood, bamboo, vegetable fibre or bark, gourds, seeds, animal and reptile skins, bones, shells and clay.
- **Animal rearing:** Animals are reared for food and ceremonial rituals.
- **Hunting and gathering:** Men and boys do most of the hunting while women and children hunt small animals and gather fruit and nuts from the forest. Traditional weapons are made, for example, bows and arrows, spears and nets.
- **Arts and crafts:** There is a great variety of art and craft including paintings, sculpture, carvings, masks, bark cloth, baskets and string bags. Decorated objects include clay bowls, jars, shields and

weapons, and musical instruments. Families pass on skills to their children. All children learn the spiritual values of their clan. Boys are taught skills such as hunting, fishing and house building. Girls are taught cooking, gardening, crafts and childcare. Selected children are taught the magic, medicine and land rights of the clan.

For you to try

- List some customs and beliefs practised in your local communities.
- Practise ceremonial songs for the traditional welcome of the chiefs.

PROCESSES AND PRINCIPLES OF PLANNING AND IMPLEMENTING A PROJECT

1. Labour

Labour is the physical work a farmer and his family do to produce the food and materials they require to support themselves. In Papua New Guinea, men and women leading a subsistence life often do separate tasks. The men generally hunt and do the heaviest clearing and cultivation; women do the lighter but no less arduous tasks. Group work is common for heavy clearing, while ownership and the day-to-day work in a garden is nearly always an individual or small family task.

Changes to the traditional roles of men and women, and to the nature of work in modern society, have had an impact on the lives of people in Papua New Guinea, especially in rural areas such as the New Guinea Islands and parts of the Highland provinces.

For you to try

- Visit local gardens to observe and record the kinds of physical work families do in the garden.

2. Materials and management practices

Modern agricultural systems create highly artificial environments. Farmers plant vast populations of specific crops and in the process kill competing species. Thus, both the natural vegetation and a high proportion of the animals, including useful insects and mammals, are destroyed. Commercial agricultural systems have little diversity and it is very hard to maintain environmental stability, even for a short time. The soil is quickly depleted of nutrients and pest numbers often increase dramatically and have to be controlled.

Eventually some of these management processes will damage the environment. For example: fertilisers that are not organic replace nutrients, but they do not replace the humus that is necessary for soil structure and the continued use of pesticides eventually results in some pests becoming resilient to this method of control.

Increasing the diversity of agricultural systems and using management methods that are less damaging can overcome these problems. In gardens there are many easy rules to follow that assist in sustainable management practices.

Sustainable management practices 

Do not burn the ground: Cut the grass under wood and trees and allow it to dry on the ground. This will mulch and protect the soil until you are ready to plant crops. When grasses and leaves are dry, collect them for mulch and compost.

Crop rotation: Different kinds of crops take different kinds of foods from the soil. Because of this, we should change the place where we plant our crops after each harvest. For example: if we grow a crop of corn, we should next plant a crop that can put plant food back into the soil, such as beans.

Three main groups of crops used in rotation are:
Heavy feeders for example, corn, lettuce
Light feeders for example kaukau, taro
Nutrient givers for example, winged beans, peanuts, makuna beans.

Mixed cropping: Planting many different kinds of crops in one area at the same time. This method is practised in traditional gardens. There are many advantages to this method:

- **Crops help each other:** for example, when corn and beans are grown together, the corn shades the beans from the hot sun. The beans in return, put nitrogen in to the soil and this helps the corn to grow well. This is known as companion planting.
- **Pest and disease control:** when there is a variety of crops, it is harder for pests and diseases to spread into the whole crop.

Trees: Trees provide food, clothing and shelter. Now many people live far away from trees and lose the benefits.

Composting: In nature things die and rot. This gives soil the organic substances it needs to feed other plants. The better the compost, the higher the crop yield, and the better the soil structure.

Mulch: Dry grass (or other dry materials) placed on top of garden beds, ridges or mounds prevents the soil from drying out. Mulch protects the soil from heavy rain and slows the growth of weeds. It also helps the ground to stay cool and wet during the dry season, and provides plant nutrients for the crowing crops.

Seed selection and planting materials: To have healthy and productive crops, quality planting materials need to be used. People should save the best of their harvest for seeds and store them well.

Work the soil: When soil is turned over, air, water and nutrients are mixed within the soil. This softens the soil and allows crop roots to grow deep into the soil encouraging strong growth.

Use fallow time: As fallow time becomes shorter, due to the pressures of providing more food, it is important to plant desirable fallow crops to ensure soil conditions remain constant and do not deteriorate.

For you to try

- Invite an elder from your community, or a government officer, to talk to your class about water, forest or agricultural management practices used in your community.
- Choose one sustainable practice and apply it to your school garden.

SCHOOL AGRICULTURE PROJECT

Papua New Guinea consists of four regions, each with a very different topography and climate. Because of this, school agriculture projects will vary according to the needs and topography of schools.

Listed below are projects to consider, depending on a school's resources as well as the amount of time and effort the school is willing to commit to generate an income from specific projects.

Regions	Type of crops	Type of animals
Highlands	Growing coffee, mushrooms, vegetable farming and landscaping	Pig, sheep, chicken, cattle, honeybees and fish farming
Momase	Growing vanilla, cocoa, rice, coffee, copra, vegetable farming and landscaping	Cattle, chicken, pig and fish farming
New Guinea Islands	Growing cocoa, copra, oil palm, rice, vanilla, balsa, vegetable farming and landscaping	Chicken, pig, cattle and fish farming
Southern	Growing oil palm, cocoa, copra, rice, vanilla, rubber, vegetable farming and landscaping	Cattle, fish, chicken, pig and butterfly farming

Project 1: Plan and design a vegetable garden to generate income for the school

Vegetable gardening should be a school's first agricultural project for many reasons: low costs, quick harvests and produce can be sold at markets for cash.

1. Factors to consider before commencing a garden project

Rainfall: The rainfall pattern in Papua New Guinea varies from province to province. Both Gulf and Western Provinces receive a lot of rainfall throughout the year. Port Moresby is dry while the

Highlands, Momase and New Guinea Islands have a similar rainfall pattern throughout the year. In some parts of Papua New Guinea the rainy season lasts from November to April.

It is important to study rainfall seasons so that vegetables can be planted just before the rain comes. The rain will help plants grow and they can then be harvested quickly. Draw up a vegetable planting calendar and follow it closely so that your project is successful.

Soil type: Good soil is essential to vegetable growth. If the soil cannot supply specific nutrients, vegetable production will decline. However, if you use chicken manure and compost, these nutrients will improve the richness of the soil and crops will grow well and produce a good yield.

Crops that add nutrients to the soil are peanuts, winged bean, soya bean and snake bean. It is a good idea to plant some of these crops in the school garden just before the long school holiday at the end of the year. These will stop weeds from spoiling the garden site and when school begins the following year, the garden can be prepared by chopping up and digging these beans back into the soil. Legume plants are important because they produce nitrogen when they die and this improves soil quality.

Humidity: In a humid climate it is usually wet; people feel sticky and uncomfortable. Some plants are suited to this type of climate, especially trees. Vegetables that grow well in humid conditions are sweet potato, pumpkin and kangkong (creeping plant).

Altitude: The higher you go in the mountains, the colder it gets and this affects the types of plants that can survive and grow in this climate. For example, in Enga (2345 metres above sea level), coconuts, mangoes, sago and even pawpaw cannot survive. This is because it is very cold; the altitude is too high and humidity is too low for these types of plants. It is important to consider altitude when you choose the type of crops to plant in your school garden. Talk to local people because they are the experts and will provide excellent advice.

2. Where to make your garden

The following points should be considered when selecting your garden site:

- The vegetable garden needs to be near the school or in the school grounds.
- The land should be flat and have deep topsoil.
- There should be water near the garden.
- Ensure the soil is soft when you dig for planting.
- Fence around the garden to keep out animals and intruders.
- A hedge or trees will protect plants from wind damage and very hot conditions.

3. Garden tools

A tool is something that makes work easier. In a vegetable garden there are many different jobs to be done.

4. Pests and disease control

The most common **vegetable pests** are:

- Caterpillars.
- Cut worms.
- Grasshoppers.
- Aphis (white and patches of insects found under Chinese cabbage leaves).

Control measure:

- Hand pick and kill them.
- Use derris dust (sprinkle on leaves).
- Apply wood ash (sprinkle on leaves).

The most common **diseases** are:

- Nematodes (small worms that live in plants sharing their foods and causing disease).
- Root rots (small worms that destroy root systems in cabbage and aibika).
- Smut (small worms that produce large amounts of black powdery materials on the plant parts – corn fruit and sugar cane).

Control measure:

- Nematodes: remove affected plants and burn them. Plant another type of crop in the same area.
- Root rots: remove affected plants and burn them. Allow the area to rest for some months before using it again. There could be too much water in your soil.
- Smut: remove affected plants and burn them. Practise crop rotation.

5. Implementation

- Select a suitable garden site.
- Decide on the type of vegetables that are to be grown. Consider demand at the local market and the costs involved in running the project.
- Commence a vegetable garden diary. The diary is a reminder of all that has been achieved and what still has to be done, and when.

Example: A simple vegetable diary

Date	What we did	What we saw
15th February	We cleared the area where we are going to position our plots. We removed the grass and shrubs and heaped them for mulch. We dug up the soil and mixed in some chicken manure.	There were a lot of weeds and grasshoppers in the garden.
22nd February	Our teacher showed us how to make a seedbed. We planted cabbage and tomato seeds and watered them to keep the nursery soil moist.	The soil looked healthier than it did on Tuesday last week. It had a better texture and there were insects in it.

6. Management of seedlings

In your school garden you will probably plant one crop of vegetables, although it would be quite easy to undertake mixed cropping which would generate income all year round. Good farmers aim to produce vegetables throughout the year. To do this, they need to plan ahead. Seedlings should be planted every three to six weeks depending on the vegetable, the variety and the climate.

How to make a seedbed

1 Prepare the seedbed in a sunny place using good soil.
2. The seedbed should not be more than one metre wide.
3. Mix a little fertiliser and manure with the soil.
4. Dig the soil well until it is smooth.
5. Make the seedbed a bit higher than the ground so that extra water can drain away.
6. Make sure that the seedbed can be shaded.

When the seedbed is ready, sow (plant) the seeds:

1. Put the seeds in rows across the seedbed.
2. Cover them with fine soil.
3. Water the seedbed carefully.
4. Put mulch over the seeds until they germinate.
5. Make sure the seeds don't dry out.

7. Managing your plot

To get the best yields from your plot, you need to manage it well from the time you plant to the time you harvest. To manage your plot well, you will need to do the following:

- Replace seedlings that died when transplanted.
- If too crowded, thin out the seedlings.
- Water the seedlings.
- Weed the plot to keep it clean.
- Apply mulch to conserve water in the soil.
- Add manure to enrich the soil.
- Control pests and disease.

These are things that you should do in your garden every week if you wish to gain an income at harvest time.

8. Harvesting and storing vegetables

You must not harvest your vegetables until they are ready. If you do, they will not be fully grown and they will not taste as they should. If you harvest too late, the vegetables will be old and tough and will not fetch good prices.

M 9. Recording the harvest

Production record: write down what was harvested, when, the quantity and the quality. Complete these records carefully, especially as you are learning agriculture in school. It helps you to understand the business side of farming.

Production records can be written in table form. On a clean page in your diary, draw a table like the one shown below. Make it cover the whole page to be sure you have enough room.

Every time you harvest something from your vegetable plot, record the weight in the appropriate column together with the date. When the growing session is finished, add up the total weight of vegetables produced. This is now a complete production record.

Production record sample

Date	Tomatoes	Carrots	Corn	Cabbages	Cucumbers
12th June	1.5 kg			2 kg	
16th June		2 kg			
19th June	2.5 kg	4 kg			1 kg
21st June		3 kg			
23rd June		4.5 kg	1.5 kg		
16th July	3 kg		2.5 kg	5 kg	
20th July	1.5 kg	2.5 kg	4 kg	2 kg	2 kg
25th July				1 kg	3 kg
Totals	**8.5 kg**	**16.0 kg**	**8.0 kg**	**10.0 kg**	**6.0 kg**

Financial records show the monetary value of all you produce. This will tell you whether the school's vegetable growing enterprise makes a profit or a loss. Financial records need to be filled in at the beginning of the season. A sample financial sheet is shown below. Find out the market value of the vegetables you have harvested (that is, see what it costs to buy them in the market). Enter the value of your vegetables in the **Returns** column. Now you can complete the financial record by adding up all the costs and all the returns. See which is the greatest then subtract one from the other to find the profit or loss.

You can see that every time a farmer buys anything he or she enters it under **Costs.** Whenever a farmer sells anything, he or she enters it under **Returns.**

At the end, when the enterprise has run for a complete year, the farmer adds up both columns of figures. By subtracting one total from the other the farmer can find out how much money he or she has made.

This is his or her profit. The outcome depends on the balance between costs and returns.

Sample financial record table

Name of enterprise: Vegetable Growing							
Costs				**Returns**			
Date	**Items**	**Amount**		**Date**	**Item**	**Amount**	
Feb 10th	4 kg lime	K30					
Feb 20th	0.8 fertiliser	K16					
Feb 21st	65 g of bean seeds	K2.70					
Feb 21st	6 g cucumber seeds	K2					
	Total costs	K50.70					

The financial record

Total returns Total costs		
Profit (or loss)		

M 10. Estimating a time frame and costs involved in a vegetable project

A suggested vegetable project information and time schedule for school

Type of vegetable	Maturity period	Seeds/cuttings per hole	Estimate of expenses	Estimate of returns
Aibika	2 months	2 cuttings/holes	20.00	200.00
Banana	18 months	1 sucker/hole	20.00	300.00
Broccoli	4–5 months	1 seedling/hole	100.00	400.00
Chinese cabbages	3–4 months	1 seedling/hole	50.00	400.00
Corn	3–4 months	2 seeds/holes	40.00	400.00
Cucumber	3–4 months	2 seeds/holes	40.00	400.00
Eggplants	3–4 months	1 seedling/hole	40.00	80.00
Lettuce	4 months	1 seedling/hole	40.00	200.00
Peanut	4 months	2 seeds/holes	30.00	200.00
Pumpkin	4.5 months	2 seeds/holes	30.00	100.00
Snake bean	3 months	2 seeds/holes	40.00	100.00
Sweet potatoes	5–6 months	3 cuttings/holes	50.00	400.00
Taro	8–12 months	1 tuber/hole	20.00	300.00
Tomatoes	3 months	1 seedling/hole	50.00	100.00
Winged bean	3 months	2 seeds/holes	60.00	100.00
Yams	9–12 months	1 tuber/hole	40.00	100.00

Reference: all prices obtained from the Madang market

11. Generating income

Selling your vegetables at the market

Village markets are interesting places. They are often very lively with people arguing, buying and selling all at the same time.

In every market goods are on sale. There are sellers who offer the goods for sale. There are buyers who want to purchase goods with money. The effects of supply and demand determine price. The higher the demand, the higher the price.

To sell school produce at the market:

- Bundle your produce (beans, pitpit, taro, aibika, and pumpkin tops).
- Only the best quality produce should be taken to the market.
- Put price tags on your produce for the buyer to see.
- Carry some coins for change.

The money you earn from the market should be given to your head teacher so that he or she can open an Agriculture Account at the Bank. This account should have three people to sign (for security reasons): the head teacher, Board Chairman and your Agriculture teacher.

If the Board Chairman is away, the head teacher and your Agriculture teacher can sign the cheque for things you may need to buy from stores in town. As you deposit money into the account, over time the bank will pay you interest on the money saved.

12. Evaluation of the school vegetable project

Evaluating the school garden project is important so that further improvements can be made to generate more income.

Sample evaluation form

	Things to be improved	Future (local conditions)
1	Imported seeds were expensive	Use local seeds (beans/pumpkin/ cucumber/ corn)
2	Tools – not enough for all students	Parents are to help
3	Nursery seed boxes were not ready	Community should help in making boxes
4	Timing of vegetable picking was a bit too early	Tomatoes to be harvested when yellow
5	Proper fencing to be constructed	Local pigs to be fenced in
6	Flooding is a problem	Relocate the garden site
7	Cabbages sold very fast	Increase the price of cabbages

Better Living

Chapter Summary

In this chapter you will have an opportunity to:

- ✔ Plan practical ways to produce food for personal consumption or to generate an income.
- ✔ Work as a team to undertake specific projects to benefit the school or local community.
- ✔ Identify and evaluate goods and services provided by a range of organisations and form specific guidelines that can be applied in assessing those services to determine which best meets the needs of your community.
- ✔ Identify appropriate materials that will allow you to design, make and evaluate a product relevant to your needs.

Syllabus References

Strand: Better Living

Substrand 1: Healthy Living

Outcome: 8.2.1 Investigate and implement practical ways to produce and prepare food for personal consumption or to generate an income.

Substrand 2: Care and Management

Outcome: 8.2.2 Work collaboratively with others to select and undertake a project based on identified needs within the school or community.

Substrand 3: Wise Consumer

Outcome: 8.2.3 Evaluate those goods and services provided by a range of organisations and make informed decisions about those that best meet their needs.

Substrand 4: Making Things

Outcome: 8.2.4 Investigate the appropriateness of materials for specific purposes. Students use their imaginations to design, make and evaluate a product relevant to their needs.

WAYS TO PRODUCE FOOD

Food is a basic human need. People get food by producing it themselves or buying it from others. By producing our own food we lessen the amount of money we need to spend buying food from others. Being able to investigate and implement practical ways to produce and prepare food for personal consumption, or to generate an income, contributes to self-reliance, sustainability and better and healthy lifestyles.

1. Gardening, raising livestock, fishing

Gardening, raising livestock and fishing are three practical ways of producing your own food for personal use or to generate an income. Advantages of producing your own food are that it is:

- Cheap to produce
- Can be sold
- Fresh
- Seasonal variety
- Convenient, as it is available to be used as needed
- Creates feelings of self-esteem, self-reliance, pride and achievement.

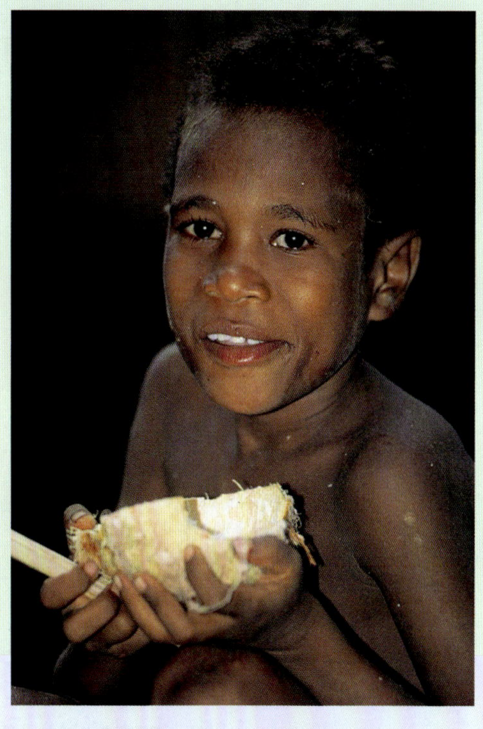

Risks to be managed include poor quality produce and loss through bad weather, pests or theft. Always use best quality plant cuttings or seeds. Make suitable security arrangements (location, fences, roofs, lighting or guards) to avoid loss.

For you to try

- Make two lists. In list one, make practical suggestions about food that your class could produce through gardening, raising livestock or fishing. In list two, name food that your class is producing for **Making a Living** activities. Compare the two lists and then discuss strategies to address the differences. Draw up an action plan with a time-line to produce food for class projects.

2. Gathering

Gathering is an informal way of obtaining food or things to help in food production. You can gather:

- Ideas for recipes
- Plant cuttings or seeds for growing food crops
- Utensils and equipment for food preparation

- Firewood for cooking
- Manure and plant matter for compost and mulching to enrich the soil in the food garden
- Vines and coconut or banana leaves to wrap and tie food parcels to cook or sell
- Grubs, nuts, ferns or bush greens to use in recipes.

For you to try

- Investigate opportunities for gathering things that help in food production.
- Discuss and put into action plans for gathering.
- Conduct a survey to gather information about food collected by your local community.

3. Processing raw materials such as kaukau

Processing raw materials can produce food. Flour can be made from sago, peanuts, beans and root vegetables such as kaukau or taro.

Kaukau flour

Ingredients

10 pieces of kaukau

Method

1. Peel kaukau and cut into very thin slices.
2. Place on a clean surface and dry in the sun.
3. When totally dry, pound the slices into flour.
4. Store in a clean dry container until ready to use.

Coconut oil

Ingredients

10 mature coconuts that have sprouted

Method

1. Scrape the flesh from the coconuts into a bowl.
2. Squeeze the cream from the flesh of the coconuts.
3. Place the coconut cream in a pot and boil for approximately one hour.
4. Scoop the fat from the surface of the oil.
5. Pour oil into clean bottles.

For you to try

- Process raw food to make a useful product. Package it to sell or keep it for future use.

WAYS TO PREPARE FOOD

1. Boiling

Boiling is a moist-heat method of cooking. Food is placed in boiling or simmering liquid over a source of heat. Using the liquid, for example, in soups and stews, is good because nutrients from the food dissolve into the liquid. However, water used to boil eggs, crabs or crayfish is thrown away.

Sago dumplings

Ingredients

2 cups sago powder
2 cups mashed ripe bananas
Cream from 1 coconut
Salt

Method

1. Mix sago and bananas in a bowl and roll into balls.
2. Place coconut cream, sprinkled with salt, in a pot and bring it to the boil.
3. Add sago balls and cook for 15 minutes.

Serving suggestion

Roll in freshly grated coconut and serve.

For you to try

- Write a recipe that involves boiling food that is available locally.
- Prepare food to sell using the boiling method.

2. Baking

Baking is a dry-heat method of cooking. Food is cooked surrounded by heat on all sides. Scones, cakes and biscuits are typical of food baked in ovens. Food may also be baked in the hot coals of a fire.

Baked sweet potato with peanuts

Ingredients

3 medium-sized sweet potatoes
1 tablespoon margarine
1 tablespoon brown sugar
¼ cup roasted peanuts

Method

1. Bake sweet potatoes in the hot coals of a fire or in an oven.
2. Mix the margarine, sugar and peanuts and cook in a pan for 10 minutes.
3. Cut sweet potatoes in half.
4. Place a tablespoon of the peanut mixture on top of each half of the sweet potato.

For you to try

- Write a recipe that uses the baking method and food that is available locally.
- Prepare food to sell using the baking method.

3. Roasting

Roasting is a method of cooking food with fat in an oven. Roast chicken and vegetables are examples of roasted food. However, pigs may be roasted on a spit, turned often and basted with fat.

Roast potato and pumpkin

Ingredients
Sweet potatoes
Pumpkin
2 tablespoons oil or dripping

Method
1. Peel and wash the sweet potatoes. Cut into serving-sized pieces.
2. Cut the pumpkin. Remove seeds. Cut into serving-sized pieces. The skin can be left on.
3. Place in a baking dish with oil or dripping and roast in the oven until golden brown. Baste during cooking.

For you to try

- Write a recipe that uses the roasting method and food that is available locally.
- Prepare food to sell using the roasting method.

4. Grilling

Grilling is a dry-heat method of cooking. Food is cooked by heat radiating from hot coals or the grill of a stove. Food needs to be turned so that it cooks evenly on all sides.

Kebabs

Ingredients
Fresh fish or meat
Capsicum or onion
Carrot
Oil
Thin sticks about 20cm long

Method
1. Cut food into small pieces.
2. Push food pieces onto thin sticks.
3. Brush with oil.
4. Cook over glowing coals or under a stove grill, turning frequently.

For you to try

- Write a recipe that uses the grilling method and food that is available locally.
- Prepare food to sell using the grilling method.

5. Smoking

Smoking is a slow, gentle, dry-heat method for cooking food in the heat of smoke rising from a fire. The smoke gives a distinctive flavour to the food.

Smoked fish

Ingredients
Small fish
Salt

Method
1. Clean fish. Fish can be left whole or filleted.
2. Place on a metal rack over a smoky fire.
3. Turn during cooking.

For you to try

- Write a recipe that uses the smoking method and food that is available locally.
- Prepare food to sell using the smoking method.

6. Mumu

Mumu is a Papua New Guinean method of cooking food in a mumu pit or hot stone ground oven. Stones are heated until they are very very hot. Using wooden tongs, the stones are arranged under, around and on top of leaf-wrapped parcels of food tied with bush vines.

Mumu-ed cassava with fish

Ingredients
4 medium-sized pieces of cassava
3 ripe bananas
3 aibika leaves
½ tin fish or flaked flesh of a small fresh fish
Salt
1 coconut
Leaves for wrapping

Method
Prepare the mumu pit until the stones are very hot. Prepare banana leaves for wrapping by softening over the hot stones, removing most of the hard stem and cutting into squares.

1. Peel and wash the cassava. Grate into a dish. Pour off any liquid.
2. Peel and slice ripe bananas.
3. Wash the aibika leaves and cut up finely.
4. Make coconut cream by scraping and squeezing the flesh from the coconut.
5. Mix together the cassava, banana, aibika, fish and coconut cream. Sprinkle with salt.
6. Place a handful of the cassava mixture onto each leaf square and wrap into a leaf parcel and tie.
7. Place parcels in the mumu pit making sure each parcel is surrounded by hot stones and leave for one hour until cooked.

For you to try

- Write a recipe that uses the mumu method and food that is available locally.
- Prepare food to sell using the mumu method.

7. Barbecuing

A barbecue uses a metal plate over an outdoor fire. The plate is oiled. Heat rising from the fire heats the plate and this cooks the food. Some barbecue frames are portable and can be moved from one place to another. Barbecued sausages, steak or chops, served with a tomato or spicy sauce, are typical of this method.

Barbecued food

Ingredients
Prawns, fish or meat
Oil

Method

1. Heat a clean barbecue plate. Spread oil thinly over the surface.
2. Using tongs, place food onto the hot barbecue plate.
3. Cook for a few minutes on one side, turn and cook for a further few minutes on the other side.

Serving suggestion
Serve with sauce, buttered bread and salad.

For you to try

- Write a recipe that uses the barbecue method and food that is available locally.
- Prepare food to sell using the barbecue method.

8. Steaming

Steaming is where the hot steam rising from boiling liquid cooks food. In Papua New Guinea it is common to place water and stones in a pot. When the water is boiling, leaf-wrapped parcels of food, such as fish and coloured vegetables, are placed on top of the stones and a lid is placed on the pot. Root vegetables may be used instead of stones. In this way, the root vegetables are cooked by boiling while the other food is steamed.

Steamed pawpaw pudding

Ingredients

1 cup mashed pawpaw
1 cup grated coconut
2 tablespoons honey
1 cup cooked rice or breadcrumbs
Juice from one lemon

Method

1. Mix all the ingredients together.
2. Wrap the mixture into a banana leaf parcel or place in a greased, covered basin.
3. Cook in steam rising from boiling water for 30 minutes. Serve with coconut cream.

For you to try

- Write a recipe that uses the steaming method and food that is available locally.
- Prepare food to sell using the steaming method.

9. Casserole

A casserole dish is an oven dish with a lid. Food (such as meat, vegetables and liquid) is cooked in the casserole dish in an oven. This is a long, slow moist-heat method of cooking. A casserole dish can be taken directly from the oven to an oven mat on the table for serving.

Coconut chicken

Ingredients

1 chicken
2 medium-sized sweet potatoes
1 small pumpkin
2 capsicums or onions
2 cups coconut cream
2 tomatoes
1 teaspoon garlic, ginger, sugar and salt

Method

1. Cut chicken into serving-sized pieces.
2. Peel and wash vegetables and cut into small pieces.
3. Place chicken and vegetables into a casserole or baking dish.
4. Add garlic, ginger, sugar, salt and coconut cream to the food.
5. Cover with a lid and simmer gently until cooked.

For you to try

- Write a recipe that uses the casserole method and food that is available locally.
- Prepare food to sell using the casserole method.

PREPARING FOOD FOR SALE

Special care must be taken in preparing food for sale so that customers are satisfied and the business makes a profit.

PD 1. Handling and packaging food

It is an offence to sell food that is unsafe to eat. People who sell food must do their best to make sure that food is safe to eat. Poor food hygiene can lead to food poisoning. Customers get sick, complain and may ask for compensation. The seller gets a bad name.

Handling food

We must keep ourselves clean when handling food so that the food is safe to eat. Keep food preparation areas very clean. We use our hands a lot, so hands must be washed before handling food or equipment used to prepare food. Always wash your hands after handling raw food, rubbish or cleaning materials. Wear clean clothing to protect food from contamination. All cuts must be covered with waterproof plasters. Hair should be tied back so that hairs do not fall into food. Rings can hide dirt and bacteria and should be removed.

Packaging food

Leaves, paper, cardboard, plastic, foil wrap and glass jars can be used to package food. Words used to describe packaging include: *see-through, thick, thin, inexpensive, rigid, soft, waterproof, protective, flexible, strong, reusable, lightweight.* Packaging can be attractive and provide information. Packaging protects food from damage. Packaging helps keep food away from dirt and bacteria. Packaging helps when handling food products during transportation and storage.

For you to try

- Make a list of rules for hygienic handling of food.
- Identify three different ways of packaging food to sell at your school.

2. Special occasions

Special occasions at your school or in the community provide opportunities for selling food. The occasion could be a school open day, cultural day, sports carnival, celebration for Independence Day or opening a new building. With lots of people present, think of a variety of hot and cold foods, on plates or in packets, as well as iceblocks and drinks that you could sell. Plan to sell food and drink at a range of prices, as people will have different amounts of money to spend. You could build a stall from which to sell the food and drinks.

For you to try

- Think of a special occasion in the near future.
- Plan food items to prepare and sell.
- Give reasons for your choice of items.
- Think creatively about ways to attract customers.

3. Number of people to be catered for

When planning to prepare food for sale, consider the number of people to be catered for. Although hundreds of people may be at your school, or at a special event, not everyone will have money to spend. If you prepare too little food for sale, you will miss out on the income you could have gained. If you prepare too much food for sale, you will have unsold food left over and you may have operated at a loss. You will need to think about the kind of food that will attract people and the amount of money different people are likely to spend.

For you to try

- Gather statistics on: the number of children in a class or school; the number of children who have money to buy food; the amount of money they have and how often they have it.
- Explain how the information gathered could help you to plan income-generating food projects.

4. Cost and equipment

Equipment such as knives, spoons, tongs, pots, bowls, stoves and refrigerators all cost money. With use over time, equipment depreciates in value and eventually it will need to be replaced. This is why the selling price for a prepared food product needs to be higher than the cost of the ingredients. To achieve a good profit margin we need to look for ways to lower production costs without lowering the quality of the product we are selling.

To save on food preparation costs, you might consider:

- Cooking with gathered firewood instead of using a kerosene or electric stove.
- Borrowing needed equipment instead of buying new equipment.

- Using equipment appropriately to avoid damage and breakages.
- Comparing prices at stores to get value for money for things that need to be bought.
- Wrapping food in banana leaves instead of plastic or foil wrap.
- Improvising equipment, for example, using bamboo tongs instead of metal tongs.

For you to try

- The ingredients to make something cost 60 toea, but you sell it for one kina. Justify the profit margin.
- Identify strategies you have used to keep production costs low in preparing food for sale. Identify how costs could have been higher if you had used different strategies.

5. Resources: human labour, materials, utensils, equipment and time management

Many resources are required when preparing food for sale. We use materials such as matches and firewood, utensils (such as bowls and spoons), equipment (such as stoves and drum ovens), human knowledge and skills, and time to plan and purchase materials. All these resources are factors to be considered when deciding on a selling price. You need to consider the hidden costs as well as the cost of the main ingredients. Common hidden costs are time, transport, power, water, packaging and equipment.

A steak sandwich, for example, requires many things in addition to a piece of steak and buttered bread. Before the customer gets the sandwich, the seller has taken the time to buy the food, transported it to the shop, refrigerated the steak until ready to cook, and purchased firewood, matches, a barbecue plate, tongs, oil, salt, sauce, a tray, foil wrap and materials to set up a selling place.

For you to try

- List resources involved in undertaking an income-generating food project. Explain how these influence the selling price of the product.
- Imagine that someone asks you to prepare a quote for making a birthday cake. Justify the price you will charge.

6. Marketing: price, product, place and promotion

To be successful you need **products** that people want to buy at prices they can afford. Commonly sold food includes: corn on the cob; cabbage rolls; cold drinks on hot days; iceblocks children can handle with ease; barbecued sausage or steak sandwiches; egg or salad sandwiches; sago and banana cakes; smoked fish; kebabs or stick food; soup; chips; popcorn; scones and doughnuts. Attractive and hygienic presentation and packaging is important to catch the attention of buyers.

Visit a marketplace to find out the **prices** other people are charging when they sell food. To be competitive, choose a similar price range. A successful food stall will have cheaper and dearer items to attract more customers. Be sure that you cover your production costs.

Be observant and notice good selling **locations**. You may have good products and competitive prices, but if you choose the wrong selling location, you will not get sales. Take time to ask advice. Walk around to select a site and get permission to use the site.

Promotion of food products for sale can be oral and written. Oral promotions can be made by radio, loud speakers, giving information in advance or calling out to people as they walk by. Plan what you want to say in a brief, polite manner. There are many different ways to advertise your products in writing. Food packages have labels stating the contents and price. Stalls can have printed lists of items for sale and their prices. Stalls often have advertising banners: 'Tasty fresh food for sale.' Printed notices can be circulated in advance. Posters can be placed in key locations.

For you to try

- Develop a marketing strategy for a food product you would like to sell. Consider price, product, place and promotion.

Care and Management

WORKING COLLABORATIVELY

Identify a need within your school or community – something that needs to be cared for or managed to contribute to better living. Working in collaboration with others, plan a project that addresses this need. The aim is to agree on what is to be done, when, where, how, why and by whom.

L 1. Effective communication skills: speaking, listening and comprehension

Shall we care for the gardens around the church?	The basketball court needs fixing. It needs a new backboard for the ring.	We could clean the beach. It is littered with driftwood and coconut husks.	I like Judy's idea of the basketball court. Lots of people use it. We could paint new lines.

Effective communication skills enable us to discuss issues harmoniously. Oral communication involves speaking, active listening and comprehension.

Speaking skills involve your choice of words, sentence order and structure, tone of voice, eye contact, gestures and facial expressions. To communicate your ideas, state your point of view concisely and give a logical reason to support it. Take turns to speak. Wait until another person finishes speaking before indicating you wish to speak in response. Avoid interrupting. Speak clearly and loudly enough for listeners to hear.

Active listening skills involve listening carefully to the views of others and then repeating part of what has been said and adding to that idea. This shows that you respect the views of others and are willing to explore them. Collaboration requires open communication and willingness to listen to and consider the views of others.

Comprehension means understanding. If you do not fully understand what another person is saying, politely ask questions to gain clarification: 'Can you explain further?' 'Excuse me, can you clarify that point?' 'Do you mean that…?' 'If I am understanding you correctly, you want us to … Is that correct?'

For you to try

A
- Form small groups to talk about things in your school or community that could improve through care and management. Appoint an observer and a recorder. The job of the recorder is not only to join in the discussion but also to record the ideas. The observer should sit outside of the group and notice the speaking, listening and body language of each person. At the end of the discussion, the observer gives feedback to the group on the effectiveness of their communication skills: how often each person spoke; any interruptions; body language of listeners; who took the lead, and so on.

2. Building good working relationships PD

Effective interpersonal skills are needed to build good working relationships. To develop effective interpersonal skills, follow these strategies:

- Be a good listener.
- Manage anger and conflict.
- Communicate accurately (verbally and non-verbally).
- Understand people's feelings.
- Develop conversation skills.
- Express feelings appropriately.
- Show care and consideration and assist others.
- Demonstrate commitment.
- Be loyal and trustworthy.
- Communicate clearly, make suggestions, obtain feedback, exchange ideas, encourage.

For you to try

- Think of a person who has excellent interpersonal skills. Analyse which actions or behaviours contribute to that person's effective working relationships with others.
- Play a team game and then identify interpersonal skills that hindered or contributed to a good working relationship between players.

3. Public relations PD

Public relations involve your relationship with the wider community. Strategies that contribute to good public relations include:

- Honest, ethical, polite, courteous behaviour.
- Respect for community property.
- Friendliness in greeting, talking and listening to community members.
- Developing ideas to improve some aspect of community life.
- Being prepared to work hard to help others and contribute to improving conditions in the community.
- Involvement in community activities.
- Cooperation in achieving results.

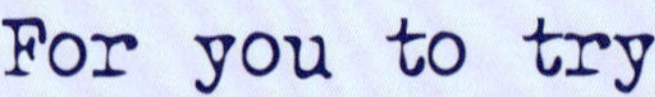

For you to try

- Plan and implement strategies to get ideas from the public about community improvement projects. Develop and distribute questionnaires. Talk about ideas with others. Interview individuals or groups. Ensure your behaviour promotes good public relations and upholds your good name and that of your school.

PD 4. Acknowledging, appreciating and respecting others' views and opinions

- To **acknowledge** is to recognise or accept the view of another person.
  Henry has suggested that we clean graffiti from signs and buildings.
- To **appreciate** is to welcome or value the view of another person.
  I thank Henry for his suggestion.
- To **respect** is to accept and not denigrate or put down a view of another.
  We respect Henry's idea. How do others feel about cleaning graffiti? Does anyone have another suggestion?

In discussion, it is important that everyone feels encouraged to contribute without fear of being shamed or degraded. All ideas should be acknowledged, appreciated and respected without value-judgments being made. Recognise, respect and value individual differences. Get everyone involved in suggesting ideas for projects.

For you to try

- Brainstorm ideas for projects that address a need in your school or community. List all ideas. Also include ideas that came from community members. Be aware of the words and actions you use to acknowledge, appreciate and respect others' views and opinions.

SELECTING AND PLANNING A PROJECT

Many factors need to be considered when selecting and planning a project based on an identified need.

1. Resources: time, cost, money, materials, tools, and facilities

Many school or community projects may be suggested. The goal of a project can only be achieved if resources are available. Ask yourself these questions:

- How many people are willing and available to do the work?
- At what time will work be done on the project?
- What costs are involved?
- Where will the money come from?
- What materials are needed? Do we have them? If not, where can we get them?
- What equipment and tools are needed? Do we have them? If not, where can we get them?
- What facilities (water, power, transport) are required? Are they available?

For you to try

- Discuss resources available for undertaking the project. Record this information.

2. Location

Location needs to be considered in selecting your project. School-based projects have an advantage because you are at that location. School care and management projects could include: making a sign board; building a new path; fencing the gardens; establishing a nursery; cutting grass; maintenance on buildings or building seats under trees.

However, if your project aims to provide a service to the wider community, to work collaboratively and build good public relations with the community, then the location of your project must be considered. Time and transport become issues when a project site is some distance from school. When planning a community project, discuss the time required to get to and from the project site and how you will transport tools, materials, equipment and people.

For you to try

- Compare the logistics of undertaking a project at school with a project in the community. Discuss the impact of location on planning the project.

3. Viability

We need to do a feasibility study to determine whether or not a project is viable, practical, possible, feasible and within our capability. There is little point in planning to repair buildings or improve a water supply if you do not have the ability or resources to achieve the desired outcome. Carry out a **SWOT** analysis. In this you identify the **Strengths**, **Weaknesses**, **Opportunities** and **Threats**.

Project options	Viable Can do	Not viable Can't do
Build new benches for market sellers		
Make a bus shelter for PMV pick-ups		
Clean and decorate the church for a special service		
Care for the paths, clean edges, fix holes		
Manage rubbish disposal arrangements		
Build a swimming pool		

For you to try

- List three projects that are viable and three projects that are not viable. Give reasons to support your opinion.

4. Tasks and activities

Identifying the activities involved in doing a task is part of the planning process. We can order activities and allocate time for each task. This helps us to plan and coordinate the people and resources required at each stage of the project and anticipate the overall time required to complete the project.

Task: to create a food garden	Week 1	Week 2	Week 3	Week 4
Prepare soil trays and plant seeds	■			
Clear the land	■	■		
Shape the garden beds		■	■	
Transplant seedlings				■

For you to try

- Identify a project goal.
  1. Create a table.
  2. List the activities involved in doing a planned project. Indicate the timeframe for each activity.
  3. Discuss who will be involved in each activity and the resources that will be required.
  4. Implement your plan. Work cooperatively with others to achieve the identified goal.

Wise Consumer

GOODS AND SERVICES: NEEDS AND WANTS

In order to make a living by producing things for personal use or to sell, we need to be wise consumers. We need to consider who provides the goods and services and whether or not we are satisfied with the quality of goods and services provided.

1. Identifying the value of goods, services and the organisations that provide them

The Wonki village truck takes people and their produce to the markets in Mendi on Saturday mornings. The fare is K10.00 per person. People are generally satisfied with the service because:

- The price is fair for the distance travelled.
- Driving is faster than walking.
- People are able to get their produce to the market to sell.
- The truck is roadworthy.
- The driver is friendly and drives safely.

The service is of value to the people because it helps them to make a living.

People value goods and services if they meet a need and are affordable, easy to access, reliable and of acceptable quality. If you want to start an income-generating project, you need to be able to identify how and why people would want to pay for the products or services you offer.

For you to try

- Read the following list of goods or services.
- Discuss the value of those you can access, and the value of the organisations that provide them.
- Write a short description of a product or service that is of value to your family. Write about the organisation that provides the service and why it is valued.

1. **Accommodation:** houses, hotels, motels, guesthouses and real-estate agents.
2. **Building:** timber merchants, concrete suppliers, building companies, hardware stores, architects and carpenters.
3. **Communication:** newspapers, television, radios, telephones and postal services.
4. **Crops and livestock:** day-old chickens, stock feed, agricultural equipment and supplies, buyers of cash crops.
5. **Education:** bookshops, suppliers of educational materials, schools, colleges and universities.
6. **Electricity:** PNG Power, generators, electricians.
7. **Financial management:** banks, savings and loans societies, accountants, and auditors.
8. **Food and household goods:** shops, supermarkets, local markets, trade stores, hardware stores.
9. **Health:** pharmacies, doctors, medical clinics, health centres and hospitals.
10. **Religion:** worship, marriage, baptism and funerals by church organisations.
11. **Safety:** police, correctional institutions, courts, lawyers, and security companies.
12. **Sanitation:** Garbage collections, PNG Waterboard, suppliers of tanks, pumps, pipes, plumbers.
13. **Transport:** PMV operators, car rental companies, trucking companies, shipping companies, airlines, service stations, tyre repair operators, mechanic workshops, earthmoving contractors, bridge and road engineers, boat yards, car dealers.

2. Importance of the services provided by organisations

Services provided by organisations are important because they supply goods and services that people need or want but cannot provide for themselves. Service organisations are many and varied.

- Air, sea and land transport companies are important because people can book and buy fares to travel to other places.
- Churches are important because they provide religious instruction and cater for spiritual needs.
- Council services are important because they maintain market places, repair roads and collect garbage.

- The Electricity Commission is important because it provides electricity services to homes and businesses.
- Hospitals are important because they care for the good health of people.
- Schools are important because they educate the young.
- Service stations are important because they provide petrol, diesel, and tyre repair and car maintenance services.
- Police are important because they help to maintain law and order in the community.
- Shops are important because they enable people to buy food, clothing and household items.

For you to try

- Identify three organisations that provide services that you use. Say why the services are important to you.
- Identify a service that your class could provide to the school or wider community and say why that service would be important.

3. Financial institutions: banking, loans, and savings accounts

Financial institutions operate to make a profit through money-related services. People value the services as a way of managing their income, savings or expenses. Common financial institutions are banks and savings and loans societies.

Many people have their salary paid directly into a bank account. Businesses deposit their income into bank accounts. Banks charge for the service and are responsible for the security of the money. This is safer than keeping large amounts of cash.

Individuals and businesses withdraw money, as they need it, to meet expenses. Excess money remains in the bank as savings. A wise consumer always has savings for special occasions or emergencies. People can earn interest on savings. Banks earn interest on the loans they provide to individuals or businesses.

Financial institutions make money by giving loans and charging interest on the money lent. Using loans, people end up paying more than the cost of the item they are buying. People value being able to get a loan if they do not have the total amount of money available at the time that it is required. They then repay the money, in small amounts, over time. Financial institutions require guarantees and can take legal action if loans are not repaid.

For you to try

- Identify ways in which financial institutions help people to make a living.
- Interview business people in your locality to find out how they use services offered by financial institutions.

4. Communication

Communication occurs through newspapers, television, radios, telephones and postal services. Communication can be used for personal purposes or to help you make a living. Different forms of communication benefit both small and large business enterprises.

Newspapers such as the *National* and *Post Courier* provide information about local and international events as well as information on products and services for sale. Businesses use printed words and pictures to advertise their products and services.

EMTV and satellite TV services provide televised audio-visual programs. Products and services are advertised in appealing ways using colour, catchy words, sound effects, music and images.

Radio transmissions communicate information in audio form. Radio stations can be contacted by people to advertise events, products and services to attract customers.

You could use the telephone to call a supplier to find out about the availability or cost of an item you need for an income-generating project. You could advise customers to telephone you to place an order for a product or service you provide. The yellow pages of the Telephone Directory contain hundreds of advertisements for goods and services.

Post offices provide services that meet the needs of customers and businesses. You can write a letter to obtain information from a supplier or to respond to a customer.

For you to try

- Consider the role of communication in helping people to make a living.
- Describe how a communication service could help you with a specific income-generating project, for example, raising chickens.

5. Electricity

People value a reliable electricity supply. Electricity is easy to use and it provides lighting and power. Electricity improves the quality of our home and community environments. Electrical goods can be purchased in stores. Electrical goods include television sets, electric fans, electric jugs, radios and music players, refrigerators, stoves and washing machines.

However, electricity costs money and needs to be used efficiently so that we can afford to pay our bills. If bills are not paid, the power supply is disconnected. People can be electrocuted, and die, so electrical goods must be in good working order and used in a safe way.

Electricity is supplied by the PNG Power Company in urban centres. Many businesses have back-up generators to use when power failures occur. In rural areas people can buy and use generators that are fuelled by diesel, petrol or kerosene. Portable generators are available and these can be moved from place to place as needed.

For you to try

- Discuss how electricity benefits the running of a business in your community. Examine how the electricity is used and the costs involved.

6. Water

People value a convenient and reliable supply of clean water. We need water for washing, drinking and cooking. We need access to clean water for good health.

Edu Ranu is a company that supplies piped water in Port Moresby. The PNG Waterboard provides piped water to businesses and residences in many other urban centres throughout Papua New Guinea. These companies ensure high-quality water that is safe to use.

Where there is no company to provide a water supply, people rely on rainwater tanks or pumps to get water from wells, bores or rivers. Many commercial businesses operate to supply moulded polythene or galvanised steel tanks, tank stands, pipelines, pumps, water filters, domestic purifiers, water treatment chemicals and irrigation systems.

For you to try

- Discuss the benefit of a reliable, clean water supply to a business in your community. Examine the water-supply system, how the water is used and the costs involved. Identify any problems with the local water supply and suggest how they could be solved.

7. Transport, police, traffic departments

Communities need transport for people and cargo. Air, sea and land transport companies meet travel needs.

Big and small aircraft and helicopters provide air services. Different shipping companies cater for sea travel, sea freight and shipping containers. Numerous trucking companies and Public Motor Vehicles (PMVs) carry people and cargo by road. There are car rental companies for people to hire vehicles. Many businesses support transport by supplying spare parts, maintenance, tyres and tyre-repair services, fuel supply, servicing of vehicles, panel beating and respraying.

It is the role of the Royal Papua New Guinea Constabulary to ensure that the public can travel safely. Traffic departments issue licences to qualified drivers and register roadworthy vehicles. Police regularly stop and check vehicles. Fines are imposed, and vehicles impounded, if problems are found.

For you to try

- Think of questions to ask the owner of a Public Motor Vehicle (PMV) to find out about his or her business operations. Consider the information you would need to know if you were to set up a PMV service.

8. The role of organisations

Let us consider the role of organisations such as Consumer Affairs Councils, traffic departments, town councils and health inspectors.

The role of the **Consumer Affairs Councils** is to ensure that people get a fair deal from traders. Their work is to handle customer complaints and to see that prices are reasonable. The also ensure that the quality of products and services is good. Consumer Affairs Councils have the power to take legal action against traders who sell poor quality products or charge unreasonable prices. People can complain to Consumer Affairs Councils if they are dissatisfied with the goods and services they purchase.

The role of the **traffic department** is to issue driving licences to qualified drivers and to register roadworthy vehicles. Special rules apply to Public Motor Vehicle (PMV) operators, as they are responsible for the safe travel of members of the public. There are severe penalties if police find unlicenced drivers, unroadworthy vehicles or unregistered vehicles on the road. They are a danger to the general public.

An important role of **town councils** and **health inspectors** is to ensure a healthy environment in towns. They look after market places where people sell food, clothing and artefacts. They ensure that there are rubbish bins and rubbish collection services. They check places that sell cooked food to ensure that preparation methods and facilities are of a high standard. Health inspectors have the power to impose fines or close shops with unhygienic premises.

For you to try

- Find out from your town council the work it does in providing services for the community. Discuss the advantages of the services provided.

METHODS OF EVALUATION

When we evaluate, we collect evidence and make value judgments based on that evidence. Evidence can be collected in different ways, for example, by observation, questionnaire, interview or document analysis.

1. Compare a range of organisations that provide goods and services

When there is more than one provider, how do consumers decide on their preferred shop, bus, bank, hardware store, church, school or service station? It could be that the staff is friendly, fast and reliable service or a convenient location.

We can compare a range of organisations that provide similar goods and services by observing the popularity of businesses, by administering a questionnaire, by interviewing customers or by reading advertisements in newspapers or telephone directories. Whichever method we use, we need to think about exactly what it is that we want from a provider.

For you to try

- Identify three shops or stores in your locality. Using the following table, give a mark out of ten for the criteria listed. Add your scores. Compare your results with those of others. Make value judgments about the services provided based on your findings.

Criteria	Shop 1	Shop 2	Shop 3
Friendly workers			
Quick service			
Clean premises			
Good supply of stock			
Fresh or safe products			
Attractive presentation			
Easy to access, convenient hours of operation			
Reasonable prices			
Accurate calculations and change given			
No long queues, no pushing, no overcrowding			

- You could do a similar exercise to evaluate PMV providers, banks or other organisations in your community. It is worth learning to select criteria for evaluating services provided. You need to consider these criteria if evaluating a business you might operate.

PD ROLES AND RESPONSIBILITIES OF CONSUMER ORGANISATIONS

1. Government and non-government organisations at the district, provincial and national levels

Roles and responsibilities of government and non-government consumer organisations are to protect and enforce the rights of people who pay for goods and services. This applies at district, provincial and national levels. Four basic rights of consumers are: the right to safety, the right to be informed, the right to choose and the right to be heard.

Safety

Goods for sale should not present an undue risk of physical harm to consumers and their families. Goods that cause harm include bad food, impure kerosene, defective vehicles, unsafe electrical appliances and drugs with harmful side effects. Police, health inspectors and the Consumer Affairs Council have the authority to enforce safety standards for consumers. They can seize unsafe goods and prosecute businesses that violate safety laws and standards.

Being informed

Governments recognise the importance of people being informed. Advertisements and labels on products should be clear and accurate. Customers have a right to be clearly informed about the contents and price of items. Many perishable packaged foods have a use-by-date to enable people to choose food in good condition. Consumer education is included in the school curriculum.

Choice

In our society we have different businesses providing similar goods and services to provide choice to customers. We can choose different sellers at markets, shops, service stations or banks. The theory is that competition keeps prices low as the sellers compete to attract customers. Where there is no competition, sellers can set the price as high as they like, up to a level where consumers will not buy the product. Consumers rely on the government to ensure that single providers of services, such as telephones, power and water, keep prices at a reasonable level.

Being heard

Consumers have a right to be heard if they wish to complain about an unsatisfactory invoice, product or service. Sometimes the business that provided the product or service will not cooperate in resolving the complaint. A consumer can take court action. Newspapers often assist dissatisfied customers by reporting unfair or unsafe products or services. Customers have a right to replacement of defective goods, a refund of money, or compensation in some form if their complaint is valid.

For you to try

- Imagine you are selling food. Discuss your roles and responsibilities for fair trading.

  How will you ensure your product is safe?

  How will you advertise and label the product to provide honest information?

  How will you offer choice to consumers?

  How will you respond if a customer has a complaint about your product?

2. Basic needs and appropriate ways of meeting these needs

As a consumer of goods and services, it is your responsibility to choose appropriate ways of meeting your basic needs.

Food is a basic need. An appropriate way of getting food is to produce your own, or buy at local markets or stores, or a combination of the two. What do you do?

People need treatment when they are sick. An appropriate way of getting better is to go to the hospital or a doctor for diagnosis and treatment. Some people use traditional remedies but these are not always reliable. What do you do?

Safety is a basic need. An appropriate way of being safe is to be security conscious. Be aware of when you are vulnerable and take precautions. Lock doors. Walk in groups. Let people know where you are so that they will raise the alarm if you go missing. Travel in roadworthy vehicles or seaworthy vessels. What do you do?

For you to try

- Describe appropriate ways of meeting your needs for clothing, shelter and one other need of your own choice.

APPROPRIATE MATERIALS

1. Quality and type of materials to use for specific purposes

Plastic, metal, cardboard, plywood, milled timber, fabric, string, cement and bricks are some of the manufactured materials available that we can use to make things. Each type of material has a range of qualities that make it appropriate to use for specific purposes.

Plastic is suitable for hoses because it is flexible. Metal is suitable for roofing because it is strong and waterproof. Cardboard is suitable for cartons because it is strong and light. Fabric is suitable for clothing because it is smooth, light and washable.

The choice of material we use for a specific purpose depends on its qualities. We use strong timber for a chair but three-ply wood to line a wall because strength is not required to the same extent. We use strong sheet metal for a barbecue plate but flexible metal wire for a cooking fork. We use strong non-absorbent nylon string for a fishing line but softer absorbent cotton thread for sewing.

For you to try

- List five things you could make, and the material you would use, to make each one. Describe the quality of the material that makes it suitable for the specified purpose. Add them to the following table.

Product	Material	Qualities
Irrigation system	Plastic tubing	Strong, flexible, light, durable
Window curtains	Fabric	Smooth, strong, washable, colourful
Furniture	Wood	Strong, rigid, able to be cut

2. Materials available in the environment

Bamboo, coconut shell and leaves, vine, shells, clay, bush timber, bones, cane, pandanus, kunai, pitpit, seeds, gourds, bark and kapok are some of the materials in the environment that we use to make things.

Bamboo bridges, coconut shell bowls, coconut leaf hats, vines to tie roofing beams together, shell jewellery, clay cooking pots, walking sticks from bush timber, bone food peelers, cane chairs, pandanus mats, kunai grass roofs, pitpit trays, seed curtains, gourd lime containers, bark for tapa cloth and kapok for stuffing pillows are just some of the materials and their uses. There are many more.

The choice of material we use for a specific purpose depends on its availability in our environment. Building materials from sago and coconut palm trees are widely used in coastal areas where these trees grow. Kunai grass for roof thatching and bamboo are widely used in highland areas. Materials found in our local area are available at low cost.

For you to try

- List ten different materials available in your local environment and give a use for each one.

3. Texture, flexibility, strength, shape

When deciding if materials are appropriate for a specific purpose, we consider qualities such as texture, flexibility, strength and shape.

Texture refers to the feel or touch of the material's surface. A tablemat could be made from tapa cloth, fabric or pandanus and each would have a different texture or feel.

Flexibility refers to the elasticity of something and the degree to which it can bend without breaking. Because we do not want flexibility in tables or bookshelves, we choose materials that are rigid or unbending. However, if we want to weave a mat or basket, or make a garden rake or strainer, we choose materials that are flexible. Pandanus, cane, vines, fabric and bamboo are examples of flexible materials.

Some material is appropriate because of its strength. Nylon string is much stronger than bush fibres and suitable for fishing line. A chair, bed or table could have wooden or metal legs. The metal legs would be stronger than the wooden legs. Bamboo, tied together to make bridges across streams, is strong enough for people to walk on.

Some material is chosen because it is smooth. Fabric, for example, is used to make clothing, bed-sheets and pillowcases because it is smooth next to our skin. Traditional clothing made from bush fibres, leaves and bark is not as smooth as fabric. In its natural state, timber is rough. We use planes and sandpaper to smooth the rough surfaces of timber furniture items so that we do not get splinters.

Shape is the form or outline of something. Bamboo tubes are suitable to make musical instruments and cooking tubes because of the hollow inner space. Coconut shells are suitable for bowls or drinking cups because of their round shape.

For you to try

- Select materials to make specific products and explain why qualities such as texture, flexibility, strength, shape or colour make the materials suitable for the chosen purpose.

4. Using appropriate technology

Technology is the name given to any system in which materials, tools or machines are used to make work easier. Examples of technology include the use of axes, spades, hammers, saws, sewing needles, water pumps, chainsaws, wheelbarrows, sewing machines, lawnmowers, stoves, kerosene lamps, bicycles, typewriters, outboard motors, tractors, paper, glass, plastic, candles, electricity, metal, telephones and vehicles.

Sustainability is important when considering the appropriate use of technologies in Papua New Guinea. This means that we must be able to maintain and continue the systems we use. Appropriate technology for Papua New Guinea must be economically, technologically and socially sustainable.

Economically sustainable technology requires low capital investment and affordable maintenance.

Sustainable technology uses local materials and energy and provides a product or service of acceptable quality, with high reliability, low maintenance and ease of access.

Socially sustainable technology:

- Uses existing skills with minimum retraining.
- Involves local people and responds to real needs.
- Creates opportunities for economic activity.
- Can be understood, controlled and maintained by people without a high level of Western-style education.
- Does not harm the environment.
- Provides work satisfaction.

For you to try

- Identify characteristics of appropriate technologies. Discuss how a technology could be evaluated to determine whether or not it is appropriate for your community.
- Identify the following items as 'appropriate' or 'not appropriate' technologies for your community: wheelbarrows, walkabout sawmills, tractors, computers, television, digital cameras, washing machines, spades, axes, water tanks.
- Add other items to your lists to show that you understand the meaning of 'appropriate technology'.

Drum oven

SPECIFIC PURPOSES

As we make things for specific purposes, it is important to create products of high quality. Here is a checklist for quality.

	There is a clear goal, that is, the expected outcome is understood.
	People have the necessary knowledge and skills.
	Necessary resources are identified and how they will be obtained.
	Each step in the construction process is clearly identified and understood.
	Materials are of good quality (available and affordable).
	Sufficient supplies of materials are available.
	Tools and equipment are available, in good condition and suited to their intended purpose.
	Knowledgeable people are around to ask for assistance if problems arise.
	The standard of work is checked at each step and re-done if required.
	People take pride in doing the work to the best of their abilities.

1. Project ideas

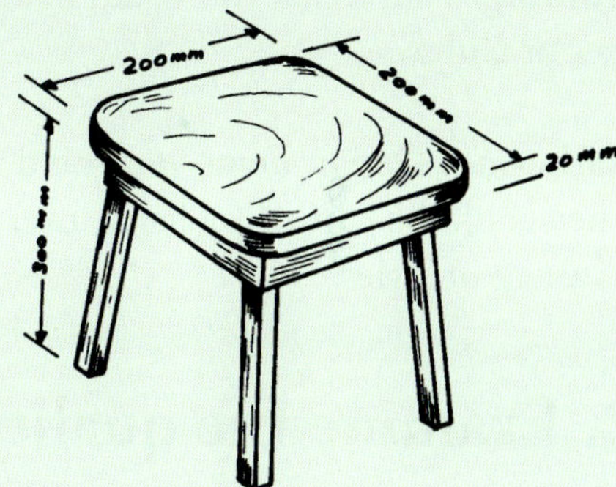

Explore project ideas to make things for specific purposes.

- Investigate the need for a range of products.
- Plan and design a range of products.
- Make or produce the items.
- Market or use the things you make.
- Evaluate the processes and the products.

Furniture: Explore alternative design ideas to make items of furniture. They could be benches, beds, chairs, cushion-chairs, tables, cupboards, bookshelves, desks or stools.

Construction: Draw a plan or scale drawing of something to construct. It could be a pit toilet, stall for selling goods, stepladder, trestle (a braced framework with a horizontal beam and sloping legs), room divider, path, footbridge or playground equipment.

Cooking: Consider things to make related to cooking. You could make a clay pot, hollow drum for mumu-ing, food safe, barbecue fireplace, coconut scraper, drum oven, tray, sawdust stove, cutlery box, grater, chopping board, tongs or cooking utensils.

Gardening: Investigate alternative design ideas for things to make for gardening purposes. They could be digging sticks, spades, garden forks, wheelbarrows, irrigation systems, strings and stakes.

Clothing: Design clothing items you could make for personal use or to sell. They could be children's clothing, dresses, blouses, skirts, laplaps, shorts, shirts, trousers, aprons or hats.

Entertainment: Create plans for items you could make for entertainment purposes. They could be flutes, drums, kundus, garamuts, ukuleles, bamboo tubes, sports equipment like bats and balls, goalposts, playground equipment or a dartboard.

Craft: Be creative in designing craft items to make. They could be photo frames, vases, mats, bilums, baskets, artefacts, walking sticks, fans, lampshades, waste-paper baskets, pencil holders or a sunshade.

Fishing: Explore ideas for items you could make related to fishing. They could be canoes, spears, traps, lines, nets or a fishing basket.

Hunting: Plan items you could make for hunting purposes. They could be bows and arrows, spears, axes or clubs.

Gathering: Design items you could make for gathering purposes. They could be bilums or baskets. They could be carried by hand, over the shoulder, on the head or on the back. Be creative in planning a variety of styles.

2. Examine the quality and quantity of materials required

Quality

We can examine the qualities of materials to understand why they are suitable for particular purposes. Materials may be strong or weak, flexible or rigid, smooth or rough, waterproof or absorbent, hard or soft, heavy or light, they can float or sink.

We know we can use:

- Strong wood for outer walls and weaker plywood for inner walls of buildings.
- Rigid wood for furniture and flexible pandanus for weaving.
- Smooth metal for cooking utensils and rough timber for fence posts.
- Waterproof sheets of iron for house roofs and absorbent fabric for clothing.
- Clear glass in windows and opaque materials for walls of toilet buildings.
- Pure clay for clay pots and a mixture of cement, sand and water for concrete.
- Hard metal for cooking trays and soft kapok for stuffing pillows.
- Heavy wood for house posts and light fibres for bilums and baskets that need to be carried.
- Material that floats for boats and metal that sinks for sinkers on fishing lines.
- Paint to cover the surface of a wood door and varnish to shine the surface of a coconut shell.

Quantity

We need to calculate the quantity of materials required for a project and check that sufficient materials are available. If we precisely calculate the quantity of material required, we can determine the costs and minimise waste. We need to make things economically. Plans and patterns enable us to estimate the amount of fabric or plywood required and the amount of waste that would result.

For you to try

Discuss and answer these questions.

- What materials would you need to make a concrete path? What qualities make those materials suitable for making a path? What quantities of each material would you need to make a path from one specified place to another at your school?
- What materials would you need to make cushions for chairs in your classroom? What qualities make those materials suitable for cushions? What quantities of each material would you need to make the cushions?

3. Finding different ways to make things

Successful creators experiment to find different ways of making a product. A skirt can be straight or gathered, long or short, with pockets or without pockets. A money wallet can have a zip, clip, flap or tie so that money does not fall out. Size and design features, construction materials and processes can be varied to create bilums, bags and baskets.

Bird house

Variety can be achieved by using different:

- Materials: bamboo, wood, fabric, metal, fibres.
- Sizes: large, medium, small, long, short.
- Shapes: round, square, rectangular.
- Finishes: paint, varnish, embroidery, appliqué.
- Skills: sew, machine, embroider, weave, knit, build, weld, paint, varnish, nail, saw, screw, carve, glue, cut, twist, file, drill, split, scrape, plane, clamp, chisel, pin, stuff, gather, fold, join, sand, measure, tack, decorate, layout, draw, rule.
- Patterns or plans: with or without pockets, with or without handles.
- Individualising work for clients: personal or organisational names or logos.
- Packaging or presentation: boxed or wrapped, in paper or plastic.

For you to try

- Experiment to find different ways to achieve your goal and to make things.
- A How many different ways can you think of making a bag, toy, seat or bowl?

Community Development

Chapter Summary

In this chapter you will have an opportunity to:

- ✔ Apply knowledge of your local community to assist the community in devising cooperative plans for economic and social benefits.
- ✔ Use effective communication skills and mediums to ensure that all stakeholders within a community are aware of specific issues that have a direct impact on the community and facilitate this awareness.
- ✔ Plan and undertake projects that will generate an income and allow you to make a living.

Syllabus References

Strand: Community Development

Substrand 1: Knowing Communities

Outcome: 8.3.1 Apply an understanding of the local community to develop and undertake a cooperative plan that provides economic and social opportunities and benefits for their community.

Substrand 2: Communication

Outcome: 8.3.2 Apply effective communication skills and mediums to facilitate awareness of an issue of concern to the community.

Substrand 3: Community Projects

Outcome: 8.3.3 Plan and undertake an enterprising project to enable them to make a living.

Knowing Your Communities

UNDERSTANDING AND KNOWLEDGE OF LOCAL COMMUNITIES

1. Culture, traditions and values PD

Culture is the word we use to describe our customs and traditions, and our way of life. Papua New Guinea has a rich and varied cultural tradition expressed in more than 800 languages, and ceremonies and rituals, songs and dances. Unfortunately, our rich tradition is under threat. Papua New Guinea is changing rapidly and our traditional way of life is under threat.

Most Papua New Guineans come from a village. The cultural traditions of our village can remain powerful influences throughout our lives. It is possible for well-educated Papua New Guineans to successfully move between traditional customs and modern ways of living. People can attend church regularly, but they can also retain a belief in magic and traditional beliefs. People can eat in modern hotels, but they can also chew buai. And living and working well with others, care and honesty, are as important in modern society as they are in traditional village society.

Customs are the traditional ways of doing things in the community. Customs have guided people for thousands of years. They are not written down in a book, but held in memory and passed down by word of mouth. Village elders are safe keepers of our customs and traditions. They also understand how customs influence most aspects of our lives: naming babies; relations between different clans; gardening; hunting and fishing; making feasts; and the treatment of sick people.

Values are the moral standards considered by a community to be important in life. But what may be regarded as important in some communities may not be so important in other communities. Our values are reflected in the beliefs and standards that guide our decisions and actions. We express our values in the way we think and act. Many clans have a special animal, bird or plant as their symbol or totem. Elders hold sacred knowledge of hunting, gardening, making clothes and bilas, and rituals and ceremonies, in trust for future generations.

In Papua New Guinea today, values are changing fast. Money has become highly valued, and our decisions and values are sometimes guided by money. Earning money links with traditional customs when we perform a singsing for tourists, or make traditional carvings or bilas for sale to visitors.

For you to try

- Invite a village elder to your school to explain the most important customs or ceremonies for your local area.

- Make a head-dress or necklace that you can use in a ceremony.
- Using pictures from old newspapers and magazines, make a poster showing the different cultures of people in the four regions of PNG. Write a short story describing each picture.
- Discuss and list at least five things that you think are important to you and your community's values. Compare your list with others, and discuss the differences and similarities.
- Invite a police officer to talk about why there is a general breakdown of law and order in our communities. In groups, discuss possible causes and solutions. Report your findings to the class and debate them.
- Write a birthday greeting card for someone you love and care for.

SS 2. Geographic locations

There is a great variety in the settlement sites of villages. A village may be built:

- On stilt posts over a river (like Kambarmba village on the Sepik River)
- Alongside fertile river valleys (like Kar village in the Waghi valley in WHP)
- On an volcanic island (like the Tabele community on Manam island)
- Above the sea (like the Motuan villagers of Port Moresby).

Each village has been selected for its particular advantages. They are located in different sites for different reasons:

- Defence from enemy attacks
- Good garden land
- Fresh water for drinking
- River transport
- Forests for building materials and food.

However, a good location does not protect a village from natural disaster:

- Rain can cause flooding
- Rivers can burst their banks
- Volcanic islands can be unstable
- Tidal waves can be a threat to those living along the coast.

Villages remain in their original location when the members of the community can benefit from the local environment and at the same time, overcome whatever natural disasters may strike the village from time to time.

Today, an improved communication and transport network makes it easy for people to move around the country. Many move from rural locations to urban areas. Village life is often thought to be difficult, and the attraction of urban life can be hard to resist: to find paid jobs; to have easy access to health and education; and more entertainment and social activities.

For you to try

- Invite a village elder or community member to give reasons why people chose your village's location.
- Walk around your community and describe the location. List some good and bad things about its location.
- List three important services provided by the government or NGOs that benefits your community.

Type of services	Examples
Social	churches—helping needy people, providing food, education
Economic	palm plantation companies—cash crops
Infrastructure	PNG government—road building

3. Family clans  

Papua New Guinea is rich in family life. Most families are extended to include many relatives. In Simbu, as in all other cultural groups, the family is the community. In Western countries, families are smaller. Many Western families include only the parents and perhaps two children – this is what is known as a 'nuclear' family. The idea of a nuclear family is foreign to most Papua New Guineans.

Many changes have affected PNG families, but some traditional ways of doing things stay the same – a family event such as a birthday, for example. Fancy wrappings and gifts of toys are unnecessary. A small cake can be enough. The important thing is for the extended family to be together, for relatives to attend in large numbers, to enjoy the food and celebration.

In PNG, we tend to live in groups and our loyalty may not extend beyond the village or the collection of houses that make up our community. Members of a ***clan*** are all descended from the same ancestors. Although clan members may live some distance apart, they remain loyal and helpful to each other. These extended families work together and share what they have as part of the ***wantok*** system.

Urban development and change has not affected the ***wantok*** system, because people in towns and settlements prefer to live together in their own ethnic group. People from the same districts, provinces, language groups and regions get together to support each other for marriages, deaths or birthday celebrations. That is how the ***wantok*** system is practiced in urban communities.

For you to try

- Explain why family is important.
- What kind of family do most Papua New Guineans enjoy? Explain.
- Debate the topic: The 'Wantok system' should not be practised in PNG?
- List important values that you have learnt from your family members.
- Name some important family rituals that are celebrated. Pick one of these and tell the other students about it.
- Draw a diagram of your extended family tree.

SS 4. Populations

A population is the total number of people living in a particular area.

The number of people living in PNG has greatly increased since the 1990s. In 1990, PNG's total population was 3.5 million. In the census in the year 2000, the population was 5.2 million. Our population is likely to continue to grow. This will mean that basic services like health and education, agricultural assistance and security, for all people, will need to be extended by the government. Service organisations like NGOs and churches assist the government to provide basic services throughout the country. Sometimes, private companies will provide basic services and users pay for them.

Increased population will also put pressure on land use. Traditionally, the sustainable practice of shifting cultivation so that garden land is left fallow for a period of 5–15 years will be reduced. This will affect soil fertility and may result in low yields and poor quality produce. If it is hard to access basic services, some people move from remote rural communities to urban areas.

For you to try

- Carry out a population census (count) of your school community. How many are males and how many are females? Record your survey as a graph.
- Discuss and list some ways to control and manage population increase.
- Class discussion topic: Why do you think that in the past most Papua New Guineans wanted many children? Write the results of your discussion in your journal.
- List some likely effects of population increase in our country.
- Discuss how population increases affect shifting cultivation method of farming?

5. How development benefits the community SS

For communities to develop, they must have services, such as health and education, good roads and bridges, access to telecommunications, churches and schools, and opportunities for people to gain employment. When good community development takes place, it greatly improves the quality of life for the people. It enables people to fulfil their needs for shelter, spiritual development, medical care, family life, peace and harmony, population growth, family planning, water, food, communication, money and education.

- A school means children can be enrolled and have a chance to be educated. It will increase the literacy level of the community.
- A health centre means everyone can get treatment quickly when they need it. This means more people enjoy healthier and longer lives and there are fewer infant deaths.
- PMV services mean that you can travel to other places. Important services are easier to reach.
- A church means that you can hear the word of God and develop spiritually.
- Connection to an electric power supply allows people to have lighting and electrical appliances.
- A water well gives families access to clean water for cooking and drinking.
- Living in or near a town gives families access to big businesses and the government.

Whether in a village or a town, people's needs must be met. Individuals, families, groups, churches and government institutions become partners in development projects. Community development can only happen when Papua New Guineans establish community networks. Communities need to be strengthened at their own pace and in ways that suit their own cultural needs.

For you to try

- Discuss and list two important effects of the following developments in your communities:
  1. Improved road networks
  2. Schools
  3. Bridges
  4. Churches
  5. Sports fields.

6. Economic opportunities

To live in modern PNG, people need money. There are council garbage collection taxes to be paid. Money is also needed to buy clothes, axes, radios, knives, plates, and cooking things. It costs money to travel by PMV or plane. When children go away to school, the family has to pay fees. In the towns we have to pay rent for the houses we live in, and food costs are more expensive.

Opportunities are opening up for people to earn money to pay for goods and services. An increase in goods and services usually means a more comfortable life and a better quality of life. Getting a formal job as a clerk, security guard, shopkeeper, teacher, doctor or PMV driver, is possible for men and women. People can find all sorts of jobs, like selling ice blocks, mending shoes, selling second-hand clothes or selling cooked food. Working on the land to make money selling cash crops is now very common.

All work is important. Whether you are working in an office for the government or a company, or selling goods at the market, or growing cash crops, you should take pride in whatever you do. With the money earned, you can pay for goods and services that will help give you a better quality of life.

For you to try

- Invite someone who is self-employed (such as an ice-block seller, betel-nut seller, shoemaker or PMV driver) to talk to your class about their work and how it helps them to make a living.
- There are two kinds of work that people do in our economy: **informal work** and **formal work.** What does this mean? Give five examples of each.
- Class discussion topic: Why is money important in our community today?
- Start a school garden. Choose where it will be, clear the land, prepare beds, and fence it if necessary. Make space and drainage for your crops. Grow crops and vegetables to sell to make money for your class picnic.
- In small groups, think about and suggest money-making activities that your class can do to raise money for your picnic.

7. Resources SS

Our society depends on our **natural resources**. Natural resources include water, land, forest and minerals. How we use these resources influences our health, security, economy, and well-being. One of the national goals in the preamble (introduction) to our Constitution talks about natural resources and the environment. It reads:

> National resources and environment: we declare our fourth goal to be for PNG's natural resources and environment to be conserved and used for the collective benefit of us all, and to be replenished for the benefit of future generations.

Papua New Guinean people are rich in ways that we may not be aware of.

- We have plenty of land.
- No one needs to be hungry in rural areas. There are plenty of edible fruits, plants and nuts, and animals that can be hunted.
- In our rivers, seas and lakes there are lots of fish and other sea creatures to eat.
- Timber from the forest, minerals and oil from the ground, fish from the sea, and cash crops grown in our land can be sold to other countries to bring revenue into the country.

Human beings are the most important natural resource in any community. Resources can only be utilised when people have the skills and knowledge to make use of them. People must have the skill necessary to trap animals or catch fish to feed their families. A community with skilled and knowledgeable people, who are prepared to work hard, will live well.

Time is a resource. Many people do not value time. They may not stick to agreed schedules for achieving talks and projects. We must use time carefully to make improvements in our communities.

For you to try

- Develop a timetable for a class activity that can generate income such as growing food or making items for sale using recycled items.

SUSTAINABLE DEVELOPMENT PRACTICES

By **development** we mean the changes that occur in a society as it grows. **Sustainable development** refers to the changes that can be sustained or maintained to grow and improve society. But not all development leads to a better way of life for the people. PNG has experienced a lot of change since Independence. Changes and development in PNG has had enormous impact on our people, in every aspect of life.

1. Efficient use and management of resources

The population of PNG is growing rapidly and this growth imposes a strain on the whole community. An increase in the number of people means an increased demand for food, water, fuel and other resources. Forests may be cleared to provide more land for building and agriculture, but too much land clearing can damage plants and animals and harm the living standards of those who depend on the natural forest.

We must look after our natural resources. Our living must be environmentally friendly and sustainable. We must manage our resources wisely and plan for the future.

Change and development challenge the values of a community. How a community participates in and manages change is vital to the community's healthy development. Ingredients of healthy development are:

- Human values
- Sustainable living
- Love of nature and the environment
- Harmonious community living and cooperation.

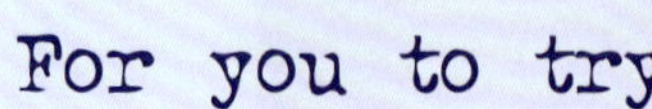

For you to try

- Invite an Environment Officer to your school to explain how the natural environment is managed in your province.
- In groups discuss how a natural resource can be used effectively to minimise damage to our environment. (Choose any one of the following resources: land, forest, coral reefs, or sea.)

- Create an awareness campaign about an important local environmental issue. Talk to other students and teachers in your school or at a P&C meeting. Prepare and use aids like charts, posters and pictures.

2. Conservation and preservation of environments  

Conservation of resources means using resources wisely to prevent them being wasted. Conservation was always important in our traditional way of life. People developed ways of looking after the soil. Traditional ways of hunting and fishing gave wildlife more opportunity to breed and multiply. Traditional practices have sustained our resources over thousands of years. Today we have to actively protect and conserve resources.

Some practical ways to conserve resources are given below:

- Use the same garden land without cutting down more forest.
- Grow trees for firewood on grassland instead of cutting down forests.
- Grow trees for houses and furniture.
- Do not light grass fires.
- Grow trees and plants that will attract wildlife such as birds and butterflies.
- Recycle old things to make other things.

By conserving our resources, we will be able to produce what we need for a long time into the future. It is important that we conserve resources for the future, if we are to satisfy our needs and improve our lifestyle in the long term. Conservation is important if our children and their children are to live well. Our resources are limited and must be managed carefully.

For you to try

- Do something practical. Set up a mini-botanical garden at your school community. Grow trees and plants which will attract wildlife such as birds and butterflies.
- Plant fruit trees along the school boundary to provide shade and fruit for your school community or your home community.
- Make a school garden or, if you already have one, carry out mulching and make composts to improve soil richness and improve production.

3. Care of environment 

Traditionally in Papua New Guinea, rural communities have a love of nature and their environment. The natural environment adds meaning to traditional life. The natural environment is their source of food, an important social and cultural site for initiation and worship, and a hiding place for spirits of dead ancestors. It is also a place for sport and recreation for young people.

Changes in values that have come with development can damage the environment. Communities and companies, greedy to exploit resources for money, can cause lasting damage. Land and forests, rivers, reefs and seas can be destroyed by pollution and reckless exploitation.

Development must be 'environmentally friendly' to protect our natural resources. Care must be taken not to harm or pollute land and water so that the natural environment may be sustained for the benefits of future generations.

A damaged Earth will not grow food. Polluted seas will not breed fish. Cleared forests will not protect animals. The natural environment must be looked after.

For you to try

- Draw a poster as part of an environment awareness campaign to display in your classroom.
- Organise for a guest speaker to talk to the class about environment matters. Invite people who may be working in your community such as NGO groups like World Wildlife Fund (WWF) or Greenpeace.
- Organise a community environment clean-up day: Cleanathon Day.

4. Capacity building

Training people to become good at their job and to learn lifelong skills is known as community capacity building. Capacity building will sustain development. Skilled Papua New Guineans are needed to benefit the community over the long term.

Trainers can be brought into local communities, to teach lifelong skills. Local people can be trained to run small business enterprises such as trade stores, walkabout sawmills or PMV operations. Specific skills can be taught, such as sewing, screen-printing, the preparation of nutritious meals and this will bring benefits to individuals and the community.

Future generations need to be skilled, so that individuals can live a sustainable lifestyle in their local communities. Together, these communities will build our nation.

For you to try

- Identify a business opportunity in your local community. Prepare a plan to set up a business, and consider the skills you will need. Write a business plan.

- In small groups, imagine you were to run a class canteen for a month. Discuss the skills you would need to run it successfully, and draw up a plan.

5. Sustainable farming practices SS

Subsistence farming is when farmers and their families get all the things they need to survive from their surroundings. Subsistence farming has been the way of life for most Papua New Guineans for thousands of years. More than 80 per cent of our total population practises subsistence farming. In subsistence farming people grow crops and raise animals not to sell, but to eat for themselves. Subsistence farming includes many different tasks: gardening, forestry, hunting, tool production, house building, cloth and dyestuff making, traditional medicine, ornament making and the production of ritual materials.

Sustainable farming practice in PNG can be shifting cultivation. Subsistence crops are grown in small clearings and the farmers and their families consume all that is produced. After harvesting is complete, the old garden is deserted and left to lie fallow for a period of 10–20 years so that the soil can recover – and the people 'shift' or move on to a new clearing.

This worked well when the population was smaller and the burden on the land was lighter. Today, our population is larger and many people are using land to grow cash crops to generate income. This can mean that the period of time for land to lie fallow is reduced. When garden land is used for cash cropping, new methods need to be used to sustain the land. Crop rotation and use of compost are new farming ideas to restore the soil and look after the land.

For you to try

- Define what subsistence farming is and explain how that farming method is sustainable.
- What does shifting cultivation mean? Explain why it is not practical. Research some new farming ideas that can help to retain and restore the richness of soil.
- A Make simple gardening or hunting tools such as digging sticks or fishing spears that can be used in activities or sold for money.
- A Collect renewable items and make things that can be sold to generate money for the school.

6. Peaceful relations PD

Communities are made up of groups of people who live and work together. Groups within a community must get on well with one another. The actions of one group will affect other groups. If people live well together and support each other, they will enjoy a better quality of life.

Sometimes there are different views about what should happen in the community. If the different views cannot be worked out, disagreement can become a problem. Many problems between people occur because they do not understand each other.

Understanding comes about when people communicate well with each other. Good communication and respect between people is important. Politeness and kindness, self-control and understanding, will create happier and more peaceful communities that allow difference.

For you to try

- List five examples of difference that can be worked out (or resolved): in the family; on the sports field; in the classroom or playground; in government; in the community:
- Look through newspapers to find reports of bad relations. Give suggestions on how they could be resolved.
- Write a list of class rules to show good behaviour in the classroom, on the sports ground and in the community.
- Each day we see conflict and violence in the media. Find an example in each of the following: television, books and stories, movies, entertainment and recreation, music and songs, newspapers and magazines. Discuss how seeing and reading about violence and hatred might influence us.
- What things could we do to encourage peace in PNG? What are the good and difficult things about having many different cultures and communities in PNG?

ECONOMIC AND SOCIAL OPPORTUNITIES

1. Social interactions: sport, religious ceremonies, celebrations, arts and craft

People like to spend time with others. Shared activities can bring joy and satisfaction to individuals and communities. In the past, people in the villages played traditional games, performed traditional dances to mark important ceremonies such as feasts or bride exchange. They shared their works of art and took part in initiations, funerals and ceremonial food exchanges. These social activities helped create good relationships within the community.

Individuals and groups in communities can do different things to relax and enjoy themselves. Playing sport, taking part in singsing and traditional dances, doing creative arts or craftworks, watching sports and games are common pastime activities. They create a strong and healthy community.

Sporting activities encourage friendship and goodwill. Religious gatherings encourage spiritual growth. These activities can help prevent bad behaviour that can lead to law and order problems. Communities where there is peace are said to be having a 'gutpela sindaun' or peaceful community living.

For you to try

- Organise a sports day with your neighbouring school. Plan the timetable, sports activities and set basic social rules.

2. Economic: goods and services

An increasing number of Papua New Guineans are looking for paid work – they want to earn money. More and more farmers now grow crops such as coffee, cocoa, rubber, oil palm and vanilla, to earn cash, rather than growing crops just to feed their families. This is called commercial farming.

While farmers are looking to earn money from growing and selling cash crops, other people are taking up activities like selling ice blocks, repairing shoes, selling betel nut, selling second-hand clothes, child minding, driving PMVs, or working in trade stores. Others work as teachers, nurses, doctors or government officers. These are just some of the activities people in our communities do to earn money.

To live in PNG today, people need money. Traditional items of value, such as shell money, pigs, birds of paradise feathers and carvings, are losing their place and value in our communities. The increased importance of money in people's lives is one of the main changes in Papua New Guinea over the last few decades.

For you to try

- Explain the difference between the formal (paid) and informal work. Give three examples each.
- Traditional items of value are loosing their place and value in society. Explain why this is so.
- What are some advantages of using money rather than the items of trade that were used in the past? What disadvantages are there in having money, compared to traditional items of value?
- Explain 'street hawking'. Discuss advantages and disadvantages of this method of selling.
- Walk around the market place and observe street hawking activities. Ask questions to find out how much money these people earn.

- Investigate and plan a street hawking business that can be developed in the classroom to generate money for the purchase of books in the library.

Communication

ISSUES OF CONCERN IN THE COMMUNITY

SS 1. Land disputes

Traditional owners, through customary ownership, own most of the land in Papua New Guinea. In developing large areas of land for forestry or mining, drilling for oil, building new roads and erecting power lines, land ownership becomes a problem. Many good projects in PNG cannot be developed because of land ownership disputes.

As the population increases, land is becoming scarce and disputes arise more often. Also, more land is being used for cash crops and less is available for growing food crops. Tribal fighting and arguments over land are common in highly populated areas like the Highlands. Sometimes these disputes force people to move to other areas to settle as block-holders or squatter settlers. Block-holders, for example, have settled on tea plantations in WHP and the oil palm blocks in West New Britain and Northern Province.

Squatter settlements are a common sight on unused land in urban areas that may be owned by the government or traditional landowners. This can become a problem when landowners or the government makes plans to develop these areas. The eviction exercises in Madang town and Lae are examples of land disputes and re-settlements.

For you to try

- Squatter settlements in urban areas are becoming a big problem. Identify some of these problems and suggest possible solutions. Write a letter to the newspaper or government with your suggestions.
- Suggest things the government could do to solve rural-urban movement and squatter settlement. Explain your plans in an article for the newspaper 'Viewpoint'.
- Debate the issue of compensation in Papua New Guinea.
- List examples of large-scale development projects in Papua New Guinea. List some of the social costs and social benefits.

2. Law and order PD

Law and order is a concern of modern society. Increased school dropouts and unemployed youths are involved in crimes. Changes in lifestyles and ways of thinking make crime more likely. Problems of law and order are social problems that can badly affect economic development. Tourists and foreign companies will not want to visit or do business if PNG is not safe.

We read and hear a lot about tribal fighting, raskols, rape and violent assault, armed hold-ups and burglary. Many people own guns without an official licence, especially in the Highlands, and guns are used in tribal warfare and by urban criminals. Growing, trading and using drugs is a problem in all Papua New Guinean communities. These images of life in Papua New Guinea, shown in overseas media, frighten off potential tourists and investors.

For you to try

- Carry out a survey to find out the most important law and order issues faced by members of your community. Make up a questionnaire for your class members to complete. Make a summary of your findings and report them to your class.
- What are the main reasons for the increase in law and order problems? Explain how they can be reduced.

3. Literacy and health PD

Literacy is the ability to read and write. The literacy level in most communities in PNG is low compared to other countries in the world. Adult literacy programs are carried out by NGOs and other community-based organisations in both rural and urban communities. Literacy programs help people to read and write in their local vernacular, pidgin or simple English. This is done so that they are able to communicate and access information to improve their lifestyles.

Health means feeling well and happy while you peacefully live with others in your community. In communities where people are literate, they can read community notices, newspapers, pamphlets, and TV advertisements about the nature and causes of diseases like HIV/Aids, malaria and TB.

Some of the important health issues in the community are clean water for drinking and washing, personal cleanliness and hygiene, how to properly dispose of rubbish and how to keep children happy and healthy. Facts about common diseases, their cause and treatment using traditional and modern medicine, good nutritional eating, personal and community safety, and dangerous animals and poisons, can also be communicated.

For you to try

- Talk to your friends in your classroom and find out about each student's parent literacy level. Interview and record information about the languages that both parents speak and write. Present your results to the class as an oral report.
- Assist the local community by offering to help in local literacy programs.
- Prepare a talk to the local community on health or environmental issues that are important to the community.
- People say that 'when we educate a man, we are educating an individual but when we educate a woman, we are educating a family'. Do you agree or disagree with this statement? Explain, giving examples from your local community.

4. Education

Education is important in all communities. It is essential that values, attitudes and skills are passed on. Adults and children must learn all the skills they need to live a happy life. Learning can be informal, for example, watching and practising skills like fishing, weaving a basket, gardening or hunting.

Teachers conduct formal learning in schools where children learn to read and write and to do mathematics. This can lead to further training for a job. However, jobs can be difficult to find, even when you have completed schooling, so we need to also have practical skills that we can apply to make a living. All work requires skills and all work is equally worthwhile.

Both formal and informal learning are important. Both types of learning can teach you the skills that will help you to survive. Each form of learning is a lifelong activity that can help you to improve your way of life. Working as a subsistence farmer or a fisherman can feed yourself and your family. Surplus can be sold for cash income. It is the same as someone working as a teacher who earns a fortnightly pay to buy food and clothes for him/herself and the family.

The new education curriculum in PNG stresses that schools need to ensure that students gain skills so that they can work for paid employment as well as learn skills to live peacefully in their communities.

The curriculum stresses that students need to:

- Respect their traditions
- Be skilled to live happily within their communities
- Earn their own living
- Like and respect each other
- Be willing to work together to benefit their family, community and country.

For you to try

- Invite a skilful hunter, fisherman, dressmaker, weaver or a wood carver to tell and demonstrate how they earn a living, and how they learnt those important life skills.
- A Collect materials and make one of the items demonstrated.

EFFECTIVE COMMUNICATION SKILLS

1. Keeping the community informed L

Communication is the process of passing and receiving information. Ways of passing information can be through community meetings, radio broadcasts, TV programs, newspapers, public notices and information pamphlets. Keeping the community informed about what is happening and the issues affecting them (such as health, nutrition, farming and marketing information) will help improve community living.

For you to try

- Make up a community noticeboard at your school or in your classroom. Paste up notices, information and community news that interests and concerns people. Update them regularly, at least once a week.
- In groups of four or five, practise having a meeting where one person acts as chairperson, and another as minute taker. Talk on an issue like keeping the school grounds clean of rubbish or how to minimise waste in the school.

- Prepare a talk about HIV/AIDS or an important community issue. Obtain posters and charts from the government or NGO organisation to give you accurate information. Present the talk at a community meeting.
- Make a role-play or drama about important community issues (land problem, nutrition, health issue) and perform it in your class. You need three to seven students.

PD 2. Meeting other people

The ability to meet and relate well to other people is an important social skill. Meeting and greeting people enables you to develop good and friendly relations with others within and outside your community. It is important to introduce yourself and speak clearly when meeting new people. Always appear friendly.

People like others to smile and greet them. We like to be secure and safe and welcome. We do not like to live, work and stay with people who are not friendly and treat us badly. When meeting or staying with other people, we should be polite and respectful. They will treat us in the same way. This helps us to live and work well together.

For you to try

- Role-play in pairs or threes, meeting a person for the first time and introducing yourself and your friend(s) to the stranger. The stranger can introduce him/herself in return.

PD 3. Maintain good relations

Communities are made up of groups of people who live and work together. They relate to each other. This means that the actions of some people affect others. If people live together in harmony, they will benefit from common support and enjoy a better way of life.

Sometimes there will be disagreement, when there are different views about what should happen in the community, and it becomes a problem. Many problems between people occur because they do not understand each other. Understanding comes through communicating with each other. We should always be polite and kind, listen to others and accept different ways of thinking and living.

4. Promote networking

Important social relationships in PNG include those between:

- Men and women
- Husbands and wives
- Parents and children
- Students and teachers
- Workers and employers
- Community and family rituals
- Church and community
- Government workers and community.

Making contacts and good relations with others will enable you to share resources, information, skills and ideas to develop your community. Networking is part of human relations. We all depend on each other for meeting needs. Our 'wantok' system is the traditional networking system that worked well in traditional societies. It is still part of modern society, even in urban communities.

For you to try

- Explain what networking is and why is it important for community development?
- List some possible benefits of the school community networking with health workers.

- Invite a health worker to your school to give information about important health issues. Design a poster to put on the community noticeboard about this issue.
- What is the 'wantok system'? How do the people in our communities benefit from it? List some disadvantages of the 'wantok system' in our communities.
- Identify a school improvement project (for example, levelling of land to make a sports ground). Present a written proposal to an NGO for a contribution to this project – it may be money, materials or a person with expertise to help in the planning stage.

ENCOURAGING AWARENESS

1. Communicating in different ways

It is easier than it was in the past, to obtain information in printed and electronic form. Information and communication technologies (ICT) have had an extraordinary and widespread impact, through radio, TV, the internet, fax, e-mail, newspapers and telephones.

Papua New Guinea needs to keep pace with information and communication technology. The more accessible information becomes, the better chance the country has of reducing poverty, addressing health issues and improving the country's economic success.

From the media and group meetings we learn about community responsibilities, care of essential government services, business activities, health issues, and literacy programs. New information and knowledge can help the people and communities to make decisions, take up new opportunities and improve standards of living.

For you to try

- Plan and deliver an awareness campaign to educate the community on how it can care for services such as a primary or elementary school. Address the school P&C meeting.
- Make a poster using information from newspapers, magazines, and information pamphlets to raise awareness of business opportunities, new farming methods, health issues, education awareness, environment conservation, and so on.

2. Effective communication skills

To benefit from information and communication technology it is important to know how to use it well. If you are talking, it is important to speak clearly and maintain good eye contact. Never mumble or look around when making one-on-one contact or addressing a group. In telephone conversations, be friendly and polite. Do not interrupt a person when they are speaking. Wait for your turn to speak and promise to call back later if you need to.

There will be times when you need to write a letter on behalf of your community. You may need to write to a donor asking for funding application forms, or for direct funding support. You may be asked to write to the local government. Whatever the reason, the ability to write a well-structured letter that communicates your message clearly to the reader is an important skill.

Report writing is another important skill. To write a report you need accurate information about what has happened and what is likely to happen next. Keep the structure simple and to the point.

For you to try

- Write a letter addressed to the Head Teacher requesting money to buy seeds and fertilisers to be used in the class garden.
- Write a report for your class on the progress of the class garden or on a sports trip that you have made. Present both a written and oral report.
- Obtain a variety of forms from the Post Office, bank and government offices, and practise filling in these forms.
- Organise a day when members of the local community can come to your school to get help in filling out forms.

3. Promoting a participatory approach

Making decisions and plans for community development as a group makes people feel important and useful in the community. This is known as partnership development.

People will take part in their communities in useful ways only when they are involved in making decisions. It is important to hold community meeting where everyone's views and opinions are expressed and decisions are made together. Listening is a very important communication skill. We find out about community feelings and needs by listening to people.

Knowing how to communicate well using words and written materials lets people know about important community issues.

A good community development is a project where people are encouraged to join in. It brings pride and ownership of the project to the community. When people are involved throughout, from the initial stage, they are able to gain skills and experiences to maintain the project in the future.

Class Project

Imagine that your school's project proposal for a water supply was approved by one of the NGO organisations. The organisation will only provide hardware materials and a specialist volunteer. The organisation leaves all other responsibilities and the installation of the water supply project to the school. Plan how to involve the school and the nearby community to make the project a success.

Community Projects

ENTERPRISING PROJECTS

To make a living, people need to be enterprising. They need to be creative, imaginative and inventive in planning products or services that people need or want. Different types of projects could include:

- Agricultural
- Handicraft
- Hospitality and tourism
- Eco-tourism
- Walkabout sawmill
- Construction and maintenance projects.

Tourism projects bring income to Papua New Guinea

1. Agricultural projects

Agriculture is the main way people earn money in rural and urban communities in Papua New Guinea. We can sell food crops such as pumpkins, potatoes, cabbages, sugar cane, carrots and spring onions, and fruit such as bananas, pawpaw, lemons, tomatoes and pineapples. We can sell cash crops such as cocoa, coffee, tea, spices, oil palm and vanilla. We can sell eggs and animals such as pigs, chickens, goats and ducks. We can sell seafood such as fish, crabs, crayfish and prawns. These items can be sold locally or in town markets or shops.

To be successful, we need to think creatively about ways of marketing our produce. Perhaps we can try growing different crops or keeping different animals to other people. We could make an agreement to supply a hotel or guesthouse. We need to be imaginative in filling a need that is not being met by other sellers.

2. Handicraft

People in Papua New Guinea are very enterprising in making handcrafts for sale. There are traditional handcraft items such as armbands, necklaces, grass skirts, clay pots, mats, carvings, baskets and bilums. In modern times, people have been creative in making new things for tourists and local people:

- The PNG flag and names are woven into bilums.
- Clocks are inserted into carvings.
- Fans, trays and pot stands are woven from pandanus.
- Chairs are made from cane.
- Lampshades and room dividers are made from shells or seed pods.
- Fabric is made into clothes.
- Plant holders, vases, toys, picture frames and wall hangings are made from a variety of materials.

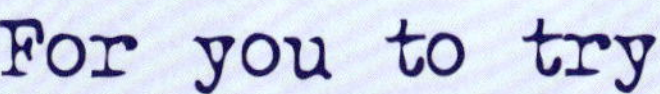

- Go for an excursion to markets and stores. List the kinds of handicrafts you see for sale. Discuss which handicrafts would be best for your school to make and sell.
- Invite skilled people from the local community to come to your school to teach children handicraft skills.
-  Make handicrafts to sell to raise funds for your school.

3. Catering and hospitality

Catering is when you provide food or entertainment for a function. It could be a staff morning tea, lunch for important visitors, traditional meals for tourists or food for someone's birthday. You need to consider: the funds available; fees you might charge; the number of people; where the function will be held; whether to have finger-food or food on plates; drinks; hot or cold food; music; dancing; resources and skills.

Hospitality is being kind and welcoming to guests. You not only provide the food and entertainment but you also act as hosts and hostesses. You could provide flower leis, welcoming speeches and small gifts. You need to be able to hold a conversation, ask questions and provide information.

For you to try

- Look out for opportunities to cater for an event in your school or community. Develop an action plan. Advertise your services. Be aware of how other groups provide catering and hospitality services.
- Consider how you could set up a guesthouse business in your community.
- Visit a guest house and ask questions from the owner: how the guest house was set up; costs; number of staff and so on.

4. Eco-tourism

Eco-tourism is concerned with attracting tourists to your community and organising activities by which they can enjoy the environment without damaging it. Think of things in your environment that visitors might enjoy.

- Could they visit a cave or a waterfall?
- Is there a beach or river where they could swim?
- Could you set up a traditional house with artefacts or craft items to sell?
- Are there special plants, birds or animals to see?
- Could they go on a canoe ride, boat trip or bush walk?

Plan tourist information to share with your class. Include information about the history and development of the place and its people. Consider warnings you give to protect the environment and for safety reasons.

For you to try

- A Write and produce a tourist brochure to advertise sights and activities in your community. Act as tourist guides and conduct a group of students to your tourist site.

5. Walkabout sawmill

A *wokabaut* sawmill is a lightweight, portable sawmill used for cutting timber. It is easy to operate and maintain. The sawmill can cut timber to be used locally and also provide income for the village people. This type of sawmill does not destroy much forest (like large-scale sawmills). Only selected timber is cut and sawn. The Papua New Guinea Department of Environment and Conservation and other Non-Government Organisations highly recommend and encourage the use of a *wokabaut* sawmill.

6. Construction and maintenance

A person with construction and maintenance skills can find many ways to generate an income in a community. Listen to people and notice the kinds of things they need or want. Produce standard items like beds, tables, chairs or coconut scrapers but also create different items like carved frames for clocks or pictures or fold-up portable coconut scrapers.

Notice structures in the community that need repair like shelters, steps, doors and market benches. Construction and maintenance activities could also include sewing or mending clothes, making curtains, cleaning houses, landscaping or providing gardening, hedge trimming or grass cutting services.

To run a successful business, it is important to study the market carefully and provide products and services that people need at prices they can afford. Advertise yourself as a fix-it person, cleaner, dressmaker, gardener, carpenter or builder on a community noticeboard.

For you to try

- Write and design an advertisement offering services such as mending, child-minding, home maintenance and place it on the community noticeboard.

PRINCIPLES OF PLANNING AND RUNNING AN ENTERPRISE

1. Self-reliance

One reason for planning and running an enterprise is to be more self-reliant. This means that we rely on ourselves to think of ways to use available resources to generate an income. By ourselves, we can get ideas from talking with people, observing how people live and reading information. We can identify our strengths and weaknesses. We can investigate resources – both material and human – to sustain a business enterprise. We have knowledge and skills that we can use creatively to plan and run an enterprise. There are advantages in being self-reliant – becoming more independent, supporting ourselves and developing pride in doing things for ourselves. People who are not getting a salary from employment need to be enterprising and make a living in other ways. This is where self-reliance is important in planning and running a business of your own.

For you to try

- Walk around your community and list activities that people do to gain an income. Ask questions about how the business helps support the family.
- Identify enterprises your class could plan and undertake to generate an income.

2. Risk taking

There are risks in starting an income-generating project. One of the greatest risks is that people may not buy the goods and services on offer. People may not need the products or services. People may not have the money. The business may be in a poor location, or other traders could be providing similar products and services at cheaper prices. If any of these things happen, the enterprise will lose money and be forced to close down.

We need to have a sense of adventure in taking the risk to start an income-generating project. We do not know if we will be successful unless we try. We need to survey the market carefully to identify a product or service that people need or want. We need to make the product in a cost-effective way and charge affordable prices. We need to consider options should problems occur.

For you to try

- Risks associated with running a chicken project could be theft, sick birds or giving credit. Identify ways to manage these risks.
- Name an income-generating project that your class could undertake. Identify a risk associated with the project and three ways to manage the risk.

3. Adapting to change

To plan and run an enterprise, people need to be able to adapt to changes when opportunity arises. It may be to plant a new cash crop, use electricity or make use of new technology. In the past, people lived in small rural villages growing food and raising animals for their own consumption, sharing extra produce with other members of the community. Today, the supply of goods and services is centred on a cash economy – money. Wants have changed from simple to more complex ones. Modern people want items such as radios, watches, television sets, cars and aeroplane travel. It is no longer possible for people to satisfy all of their wants by their own direct efforts. Instead, it is necessary to rely on individuals or groups of people who have special skills to produce those things that are needed or wanted. This is a relationship where businesses and people depend on each other.

For you to try

The diagram below shows the relationship between business and people in modern society. Discuss what business and people provide to each other.

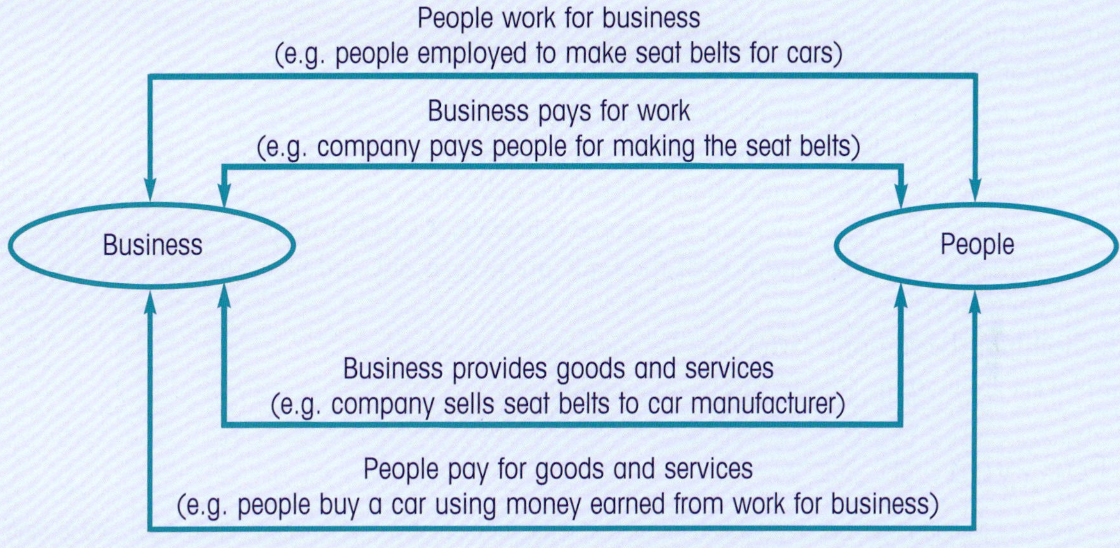

4. Solving problems

The goal of a business operator is to satisfy customers while making a profit. Some problems associated with running a business include:

- Dealing with complaints.
- People wanting credit.
- Controlling and maintaining stock.
- Funds to start a business.
- Loss of goods through theft or pests.
- Power failure for electrical equipment.
- Bookkeeping and accounting.

For you to try

- Invite a trade store owner to talk to your class about how he or she runs a business, and the problems he or she has encountered.

5. Registration of a business

Businesses operating in Papua New Guinea must be registered. In many cases, businesses are required to register with a local authority such as the provincial government or local level government. Registration forms should include the following information:

- Name of the business.
- Names of the people controlling the business.
- Location of the business.
- Details of activities carried out by the business.

Retail stores (such as trade stores, tucker shops and supermarkets) must have a licence to trade. In addition, certain items, such as meat, liquor, firearms, medicines, and insecticides, require a separate licence to sell them. Before a trader can be granted a licence, his or her premises must be inspected. This is to make sure that the condition of the trade premises is safe and meets health regulations. Inspectors check that:

- The health of the consumer who pays for food is not at risk.
- Employers are working in safe and healthy conditions.

For you to try

- Practise filling in business registration forms and other forms that you will need to fill in if you are to start a business.

6. Understanding taxation

Taxation is a system of compulsory payments by people and businesses to the government, in order to fund public spending. In deciding who should be taxed and what should be taxed, and how taxes should be levied, the government tries to keep the tax system fair. Taxes are the main source of money for the national, provincial and local governments. The government collects taxes from incomes, exports, imports and businesses. The government collects the money in the form of taxes and uses the money to fund development projects and run the country. In this way the government distributes and shares the nation's wealth among all the people. People who pay taxes are called taxpayers. Governments consider three things to work out how much each taxpayer should pay:

- What people own.
- What they spend.
- What they earn.

Some types of business activity or products, such as cigarettes, may be discouraged by heavy taxes.

For you to try

- Invite a guest speaker from the Internal Revenue Commission to give a talk:
  Who pays taxes?
  How much do they pay?
  How are taxes collected?

7. Labour laws

The government has laws to protect the workers in business activities. The government sets laws to make sure that the workers' rights are protected in the following ways:

- **Minimum wage:** The government has fixed a national minimum wage for workers. This is the lowest wage which employers may pay their workers. It applies mainly to unskilled workers.
- **Overtime wage rates:** The normal number of hours worked by employees each week in Papua New Guinea is usually 40 to 42 hours. Employees must be paid a higher rate of pay, known as overtime, if they work longer hours.
- **Holidays or recreation leave:** Workers are entitled to an annual holiday with pay, after working for the same employer for one year. Urban workers get three weeks annual leave. And rural workers get two weeks annual leave.
- **Sick leave:** Workers are entitled to sick leave with pay, if they have been with the same employer for at least three months.
- **Long service leave:** Employees who work for the same employer for 15 years are entitled to six months leave with pay.
- **Dismissal of workers:** If workers are to be dismissed, they should be given notice. A worker who has been employed by the same employer for at least three months should be given one week's notice if told to leave. Casual workers may be given one day's notice.
- **Workers Compensation Insurance:** Employers must insure their employees against injury while they are at work.

For you to try

- Invite a person working with the Labour Department to come to your school and give a talk on labour laws.

8. Quality control of products

Price control: All retail stores must obey price control regulations when deciding the price of their goods. The Price Controller's Office sets limits on the amount that sellers may charge for many types of goods. Most of these goods are essential items. For other non-essential items, the Price Controller's Office does not fix a minimum price that the retailer (or seller) must charge. Instead, it decides on the maximum mark-up a store may add to the cost of goods. This cost covers the cost price, plus the transport cost. Stores are allowed a higher mark-up price if they import directly from overseas or buy directly from the manufacturer.

Price controlled items are usually dearer in remote places. This is because transport costs are higher.

How price control is enforced? There are a number of Consumer Affairs Bureaus located in urban areas of Papua New Guinea. Their job is to check stores in their area to make sure that there is no overcharging. The prices of all goods must be displayed. If shop owners are found to be overcharging, they will be told to reduce the price. They are fined if they continue to overcharge.

The Goods Act is a law to protect consumers if they are sold goods of a poor quality. What should someone do if goods are rotten, broken or food consumption dates are overdue? They should first complain to the store where the goods were bought. They should ask for the goods to be replaced or ask for a refund of money. If the store is not willing to help, the customer should go to the local Consumer Affairs Bureau. Officers from the bureau should then investigate the matter. If a customer buys goods that are underweight or short-measured, they can complain to the Standards Division of the Department of Trade and Industry.

For you to try

- Look at how the cost of a big tin of Besta fish is calculated:
  The cost price of a tin of fish is K2.00 and transport costs are 10 toea.
  The cost into the store is therefore K2.10.
  The mark-up allowed for fish is 11%.
  $2.10 + (2.10 \times 11/100) = K2.33$

- Calculate the controlled price of the following items.

Item	Cost price	Transport cost	Allowed retail mark-up	Controlled price
Tin of meat	K3.00	10 toea	13%	
Parcel of sugar	K2.00	10 toea	11%	
Loaf of bread	K1.50	5 toea	11%	
Packet of rice	K1.80	5 toea	12%	

- Carry out a price survey. Select up to three of the big shops in town or three different trade stores in your community. Select five essential goods. Write down the prices of each of these items in each of the stores. (Try to select items that are sold in each of the stores.) Compare the cost of the individual items on your list from each store. Write down differences in prices. Suggest why this might be. This should tell you where it is cheapest to shop.

9. Money management

Most modern consumers do not have enough money to buy all the things they would like to. Have you had this experience yourself? When we are given money to spend, money earned as wages, money earned from selling our local produce or money earned in other ways, it is always possible to find many more ways of spending it than there is money available. Consumers must make decisions as to how they are going to spend their money. Planning, spending and making careful choices is one way of being sure that money will be spent on the things that are really needed. The plan of how to spend money is called a 'budget'.

The first step in planning a budget is to find out exactly how much money is available to spend. Then plan to keep your budget within that total amount of money. Also, prepare a priority list of items to buy – a list of wants, from the essential down to the least important. The money left after buying all the essentials can be kept as 'savings'.

Businesses that provide goods and services must also plan for the future. Usually they try to plan for one year ahead. They must try to guess how much income they will receive in one year, and they must also plan the way in which they are going to spend this income.

- How much on wages?
- How much on raw materials?
- How much on electricity and water?
- How much on repairs and building improvements?

For you to try

- Ask your school principal to talk about school budgets. How much is given to the school by the government? How does the school budget?
- Practise filling in forms relating to banking, for example, opening an account.

10. Sustainability

One major concern in business is how to sustain business activities or production. Business must keep operating during bad times, when demand is low. When natural disasters strike, when raw materials are scarce or too costly, when there is recession and people cannot afford to buy, how can businesses survive?

Here are some ways to encourage sustainable business activities:

- Good money management – budgeting money, saving and investing profits in the good times that can be used when bad times occur.
- Good human resources – skilled and hardworking workers.
- Self-reliance – business activities or production to be based on local production and resources.

For you to try

- Identify sustainable business practices that your class could do to run a successful business that will generate profit. For example: carry out good market research; write a financial plan and budget.

Project sheets

Project 1

DRUM SEATS

Use

Drum seats are made out of empty 40-litre fuel drums. Drum seats can be used as outdoor seats. They last much longer than wooden outdoor seats.

Materials needed

Half-cut, empty 40-litre fuel drums
River stones and sand
Cement

Tools needed

Chisel
Mallet or a hammer
Bucket
Spade or shovel

Steps

1. Collect empty 40-litre fuel drums. Use a cold chisel to cut the drums in half. (Do not remove the top or bottom lids.)
2. Collect river stones and sand. The stones will be used to fill up the drum.
3. Select a site for the seat and position the half drum. Make sure the end with the lid is on the ground.
4. Fill the empty drum with stones. Leave a 3 cm space for concreting.
5. Fill the space left at the top with sand. Pour cement on top and use water to smooth and level the surface. Leave to dry.

Project 2

POT PLANT PLATORM

Use

This pot plant platform is used for plant pots to sit on in a family's nursery or greenhouse.

Materials needed

4 pieces of post (bush timber or sawn timber) 50 cm height is ideal but can be 60 cm
2 pieces of rail support (bush timber or sawn timber) 30 cm in length
2 pieces of pot plant seats: 3×1 sawn timber or 1 piece of 6×1 sawn timber
Bush ropes and nails (3 inch nails)

Tools needed

Saws
Hammers

Steps

1. Collect all the materials needed.
2. Measure and cut the post to the required height. If the post is bush timber, sharpen the ends. Plant the posts firmly into the ground.
3. Nail and fasten the rail supports onto the posts.
4. Nail and fasten the seats onto the posts. If you are using 3×1 timber, leave a space between pieces of timber to allow water to flow out from the bottom of the plant pots. If using 6×1 timber, no space is needed.
5. Put your pot plants or seedling boxes onto the platform.

Project 3

FOLD-UP PITPIT FRUIT DISH

Use

This dish is used to hold things such as fruit. The advantage of this style of dish is that it folds up and is easy to store or carry.

Materials needed

32 × 30 cm lengths of thin pitpit
4 × 38 cm lengths of thin pitpit
Sandpaper
Varnish
Tie wire

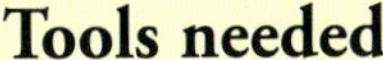

Tools needed

Saw
Long-nosed pliers
Paintbrush

Steps

1. Collect pitpit stalks, remove leaves and smooth with sandpaper. Cut 32 sticks, 30 cm in length. Cut 4 longer sticks 38 cm in length.
2. Place side by side in order, 4 short sticks, 1 long stick, 8 short sticks, 1 long stick and 4 short sticks. This is one side of the dish. Repeat for the second side of the dish.
3. Thread wire through the 18 pitpit sticks, 3 cm from the end, to join them together at one end. The wire end can be sharpened and heated to pass easily through the pitpit. Use pliers to cut and bend the wire ends to make loops at both ends. Repeat with the sticks for the second side of the dish.
4. Place one side on top of the other side and interlace the base of the dish by alternating one stick from one side with a stick from the other side.
5. Thread wire through the 36 pitpit sticks 3 cm from the end (11 cm for longer sticks) to join them together. Use pliers to cut and bend and make loops at both ends of the wire. The dish should now be able to be opened to make a dish on a stand and to be folded flat when stored. The four longer pieces of pitpit provide the stand for the dish.
6. Finish by varnishing the sticks to give a shine to the pitpit.

Project 4

HAND-HELD FAN

Use

Fans are used to keep people cool.
Fans come in different styles.
A variety of local materials can be used.

Materials needed

Pandanus leaf strips
Tapa cloth
Old ruler or stick 30 cm long for handle

Tools needed

Small sharp knife for cutting
Sewing machine and thread

Steps

1. Make a paper pattern for the fan. Draw the outline of the shape you would like on a sheet of newspaper. Most fans are about 30 cm in width and length.
2. Lay overlapping short strips of pandanus leaf in a circle. Place the paper pattern on top and cut the leaves to the shape of the pattern. Machine stitch to hold together.
3. Cut two pieces of tapa cloth to cover the inside ends of the leaves. Stitch to both sides of the fan. Leave an opening at the bottom for the handle to be inserted.
4. Use an old ruler or a ruler sized stick for the handle. Using thin strips of pandanus, weave a cover over two-thirds of its length. Insert the uncovered end between the two layers of tapa and secure the handle to the fan.
5. Finish by tying a loop of pandanus at the end of the handle to carry the fan.

NOTE: Experiment with other designs using materials available in your local area.

Project 5

RAISING CHICKENS

Use

For consumption and sale.

Materials needed

20 day-old chickens
A local bush material house
3 watering containers
3 food containers
Sawdust
20 litre kerosine
3 hurricane lamps

Equipment needed

5 bags of starter (chicken food)
3 bags of finisher (chicken food)
A brooder (small box 2 m × 1 m)
A spade (for turning the sawdust)

Steps

1. After building your house, add sawdust (5 bags).
2. Place a brooder in the middle of the building.
3. Light the lamps to keep the house warm.
4. Food (starter) and water needs to be provided.
5. Open the two boxes and allow the young chicks out on to the floor of the new building.
6. Put the young chicks into the brooder with water, food and a lamp (for the night).
7. Make a hole above the brooder for ventilation.
8. Let the chicken out early in the morning. (Add fresh water and food.)
9. Week 5: feed your chicken with finisher (chicken food).
10. Sell your chickens from week 8.

NOTE: Feed your chickens with starter from one-day-old to week 5.
Start feeding your chickens with finisher from week 6 onward.

Project 6

BUILDING A FISH POND

Use
The fish pond is for growing fresh fish for school consumption and sale at the local market.

Materials needed
Any waste land or swamp area
50 young golden carps (freshwater fish)
Spades, hoes and digging forks
String and a metre ruler

Steps
1. Measure the area (10 m × 5 m) and 4 m deep.
2. Dig the area with spades, hoes and forks.
3. Level the bottom and sides of the pond.
4. Add fresh water into the pond.
5. Add 6 buckets of compost soil into the pond. (Fish will love it!)
6. Gently place the fish into the pond.
7. Feed the fish with cooked kaukau and worms.
8. Grow water lilies in the pond as shade for the fish.
9. It is important to regularly clean the pond.
10. Harvest the fish after 10–12 months.

NOTE: Harvest only the larger fish for consumption and sale.

8 Additional Material: Projects and Investigations

MANAGING RESOURCES

1. Rainfall

- Using the library, read about the rainfall patterns of other provinces.
- Class discuss why Gulf, Western and Baimuru received a lot of rainfall throughout the year.
- Students draw the rainfall pattern for their province.

2. Soil and compost

- Students examine the topsoil in their garden plots. List in their books what they saw.
- Class construct a compost heap for their plots.
- Using the library, students study the nitrogen cycle and draw the diagram in their books.

3. Humidity

- Using the library, students read about humidity and which plants grow best in this weather condition.
- Study the map of Papua New Guinea and calculate the average humidity of your province.
- Why are farmers so concerned about plant humidity?

4. Planting in Enga

- Students locate Enga on the map and suggest what sorts of plants would best grow there.
- Make a list of vegetables that would grow successfully in Enga.
- Discuss the vegetable and livestock production options for Engan farmers.

5. Vegetable crops

- Students list root vegetables and estimate their maturity dates. They can check these dates with parents and elders in the community.
- Class discuss from the list in activity 4 above, which vegetable crop has low estimate returns. Why?
- Why is there more demand for sweet potatoes in the highland provinces than there is for taro?
- Class discuss the types of crops that are classified under the vegetable group.
- Students group these vegetables under – root vegetables, leaf vegetables and fruit vegetables. Teacher can remind the class that vegetables are short-term crops.
- Teacher and students can explore the advantages of growing vegetables.

6. Economic importance of crops and animals

- Students and teacher discuss the economical importance of crops and animals in their region and how these influence the traditional culture of the society.

7. Your school garden

- Students go out to the school garden and select a good site for a vegetable garden.
- Class draws a grasshopper and labels the different parts and describes their function.
- Discuss reasons for harvesting vegetables at the right time for sale.
- Students practise good garden management in their small vegetable plots.
- Teacher shows how to transplant seedlings from nursery beds to the garden.
- Class investigates the best organic materials for mulching the garden.

8. Selling vegetables at the market

- Class finds out the different prices of vegetables sold at the market.
- Ask local women to show how to bundle the different vegetables sold at market.
- The class carries out a market survey. Find out about price, demand for certain vegetables and whether the goods sold at the market consisted of three main food groups (energy, protein and carbohydrates).

9. Seeds and seedlings

- Discuss ways to collect local seeds and preserve them for your gardens.
- Which seeds did not germinate in the nursery boxes?
- Which insects destroyed the seedlings in the nursery? How can you control them?

BETTER LIVING

1. **Class cookhouse**
   Create different designs for a class cookhouse that is possible to build with resources available. Choose one design, develop a plan of action with a time-line. Implement the plan.
2. **Class shopping centre**
   Set up a class shopping centre. As a seller, role-play how to promote the products you sell and how to give friendly service to customers. As a shopper, role-play how to be a wise consumer.
3. **Craft book**
   Write a project sheet for a craft idea from everyone in the class. Include a diagram or photograph and instructions. Select the most suitable ideas and make them into a book that can be photocopied, bound and sold.
4. **Do-it-yourself maintenance jobs**
   Identify repair and maintenance jobs that can be done as class projects in your classroom, school or community. Advertise your services to others, at reasonable prices.
5. **Improvised cooking utensils and equipment**
   Be creative in designing and making cooking utensils and equipment from local materials and by recycling available resources.
6. **One-pot cookbook**
   Create recipes for balanced meals that are cooked in one container, for example, a meal in a coconut shell, stuffed pumpkin, sweet potato or pawpaw, fried rice or bamboo tube meals.
7. **Preserving**
   Preserve food using different methods. Make jam, pickles, smoked fish, frozen fruit juice as ice-blocks or dried fruit slices.

8. **Screen-printing**
   Set up a T-shirt screen-printing service. Advertise your designs at affordable prices.

9. **Target a tourist**
   Observe buying habits of tourists. List things that you could make to appeal to a tourist from overseas. Design and make something to sell. (Local people could buy it too, but design it for a tourist.)

10. **Toys**
    Design, make and sell a toy or gift for a child. Identify the qualities that make it safe and appealing.

11. **Campaign for a clean environment**
    Identify unsatisfactory rubbish disposal practices in your locality. Plan and implement a campaign to keep your community beautiful.

12. **Catering service**
    Plan to cater for functions such as parent or teacher meetings, birthdays or other special events. Estimate the cost per person. Design menus in different price ranges. What could you provide for K2.00 a person? K5.00 a person? Or K10.00 a person?

13. **Food stall**
    Create realistic plans for a school canteen or food stall. Choose the site, collect building materials, construct and evaluate.

14. **Food to sell**
    Plan a range of foods to sell throughout a year, considering food availability at different times of the year. Include fresh fruit, drinks, ice-blocks, ice cream, sandwiches, salads and cooked food.

15. **Furniture making**
    Create designs for furniture items that could be made from new or recycled materials in your community. Make and sell a variety of furniture items. Evaluate designs and consider ways by which they could be improved.

16. **Jam or pickles**
    Make jam or pickles from local fruit or vegetables. Build up a collection of clean glass jars or buy cheap drinking glasses. Design labels. Cover tops with a circular piece of cloth held on with a rubber band.

17. **Maintenance of the community environment**
    Advertise services (voluntary labour or for a fee) to repair and maintain buildings or areas in the community environment (sports fields, roadsides, church gardens, market benches or bus stops).

18. **Marketing**
    Design a distinctive logo, motto, brand, label or packaging for your 'class business'. Create a 'jingle-type' advertisement that can be sung over the radio. Plan how to market your products using advertising strategies.

19. **Recipe book**
    Create and sell a recipe book with recipes for foods that are available in your locality.

20. **Rising prices**
    List five common items that families buy on a regular basis. Record the brand and size of each item. Choose two shops. Once a month over a year, record the prices for these items. Discuss price variations and how this might affect family budgets.

21. **Wise consumer habits**
    Plan and implement a project aimed at raising awareness of wise consumer habits.

COMMUNITY DEVELOPMENT

1. **Clothes line**
   Design a clothes line to hang and dry your wet clothes after laundry. Plant two posts not more than 6 m apart. Nail a 'T' cross on top of each post. Tie strong rope or wire rope from one post to the other post. Leave space between the ropes for drying.

2. **Classroom news board**
   Use a piece of thin plywood (1 m × 1 m) or empty cartons as a class news board. Divide the spaces into sections: World news, National news, Local news, or Health, Sports, Social news. News can be collected from newspapers and magazines and displayed weekly.

3. **Rubbish-drum stand**
   Design a rubbish-drum stand for standing the rubbish drum close to the road near your house. A stand makes it easier for the rubbish collection workers to lift the drum and dump rubbish onto their rubbish collection truck or tractor for disposal.

4. **Market food display shelves**
   At markets in rural communities in PNG, food and items for sale are most often displayed on the ground. It is unhygienic. Collect bamboo, strong pitpits, bush timber, palm barks and ropes from the bush to build market display shelves at waist height.

5. **Building stalls and shelters and stages**
   Building stalls to sell food, shelters for shade and stages for speakers and guests at community gatherings, school open days and carnivals. These rough shelters can be built from bush materials such as bamboo, bush timber, bush ropes, coconut leaves, kunai grass and other leaves and bush materials. Design, plan and build shelters.

6. **Community or school 'cleanathon'**
   Make people aware of the importance of living in a clean and healthy environment. Work with your village elders, local councillor, head teacher and teachers to organise a clean-up day. Make fences around the village or school. Clean village cemeteries, community halls, church buildings, beaches and market areas.

7. **School bell or church bell**
   Your present school bell does not give a loud sound. Many students come late to school. What materials would you collect to make a bell?

8. **Cassava or tapioca scraper**
   Tapioca cakes are a favourite food to eat at home. They are also sold in markets. Make a cassava scraper. It can also be used for scraping banana. Collect a thin sheet of metal (30 cm × 40 cm). Nail strips of timber under the edges. Make a handle at one end. Use a hammer and nail to make many little holes on the metal sheet. The rough surface is used for scraping.

9. **Tyre plant pot**
   Tyre plant pot is made from used tyres. Collect old used car tyres (small car tyres are an ideal size). Cut off the inner round edges using a very sharp knife. Press the back of the tyre inward while pulling the inner edges backward. Cut flat pieces from the rubber tube and sew and cover the bottom end using a string rope. (Sharpen one end of a wire and heat it in the fire until the end becomes red-hot. Make holes through the tyre for sewing). Fill the tyre with soil and plant your desired plant. Paint designs. It can be sold for money.

10. **Weaving mats**
    Mats are useful to sit on, sleep on and for serving food. The types and styles of weaving mats vary from region to region in PNG. Materials used for weaving mats are pandanus leaves, coconut leaves, oil palm leaves and water straws. Invite a skilled person from the local community to the school to demonstrate making mats. Students can weave mats. These could be sold in the market for money.

Acknowledgments

The authors and publishers wish to thank the following authors and copyright holders for granting permission to reproduce their material:

Crawford House Publishing Australia: pp. 2 (middle), 34 (bottom left), 72, 84 (bottom), 98 (middle), 124 (bottom left), 156 (middle, bottom left), 157, 164, 168. Barbara Hodgins: p. 34 (middle), 124 (middle). © Dr Stuart Miller: pp. 3, 120, Lochman Tranparencies; © Alex Steffe: pp. 6, 7, 18, 27, 47, 63, 67, 71, 83, 84 (top), 92 (top), 100, 103, 109, 110, 111, 125, 138, 154, 158, 162 (bottom), 163, 166, 177, Lochman Transparencies. © Jonathan Chester: Trekking group crossing rope bridge in the New Guinea Highlands, p. 180, LONELY PLANET IMAGES; © Jerry Galea: A young girl from the village of Malala, p. 78, LONELY PLANET IMAGES; © Chris Mellor: Salesman in craft shop selling tourist souvenirs, p. 179, LONELY PLANET IMAGES; © Gary Steer: Young girl takes a shower outside her home, Karkar Island, p. 16; A man in a penis gourd aiming arrow, p. 104, LONELY PLANET IMAGES. Colette Modagai: p. 112 (right). Pam Norman: pp. 191 (all), 192. PhotoDisc: pp. 14, 53, 62, 144, 145, 184. PNG National Aids Council: p. 171. Stephen Ranck: pp. 2 (top), 46 (bottom), 70 (bottom right), 86 (top & bottom), 87, 98 (top), 102 (top & bottom), 115, 142, 156 (bottom right), 159, 170, 178. Irene Sawczak: pp. 2 (bottom left), (bottom right), 5, 9, 10, 11, 13, 21 (bottom), 23, 24, 25, 26 (top & bottom), 29 (top & bottom), 30, 32 (top & bottom), 34 (middle & bottom left), 36, 38, 46 (top), 49, 54, 57, 58, 70 (top, middle & bottom right), 75, 77, 82, 92 (bottom), 93, 94, 98 (bottom left & right), 99 (top & bottom), 101, 105, 106, 112 (far left, middle top & middle bottom), 116, 117, 119, 121, 123, 124 (top & bottom right), 134, 137, 141, 151, 152, 155, 156 (top), 161, 165, 172, 181, 183, 189.

Cover: Crawford House Publishing Australia (bottom). Irene Sawczak (top & middle right). © Michael Gebicki, Local Kavieng fishermen with net, (middle right), LONELY PLANET IMAGES.

Every effort has been made to trace the original source of copyright material contained in this book. The publisher would be pleased to hear from copyright holders to rectify any errors or omissions.

The Community Game

RULES

- Each player, in turn, throws a dice and moves the designated number of spaces.

- If you land on a square with text, read the statement. Suggest why this problem is a concern and suggest how the community should deal with the problem. Example: *rubbish in creeks – rubbish will pollute the water and the water may become unsafe for drinking. The school could conduct a cleaning up day at the creek and collect and dispose of all rubbish.* NOTE: The group has to accept your answer before you continue to play.
- The winner is the first person to complete the community trail.

Glossary

alternative	instead of something else, like a crop planted instead of another crop
amenities	things that are pleasant or useful in a house or a community
ancestors	a relation who lived long ago
appropriate	suitable, a proper way of doing things
bacteria	tiny living things, that sometimes cause disease
balanced meal	a healthy meal that contains some food from each of the food groups
barter	a trading system where you exchange goods without using money
biodiversity	the variety of plants and animals in the environment
blend	to mix together
botanical	relating to plants or the study of plants
carbohydrate	a substance found in food such as yams, sago and potatoes
caterpillar	a small crawling creature that eats plants and turns into a butterfly or moth
ceremony	a special event to celebrate something, such as a yam harvest
chemical fertiliser	fertilisers made in a factory
commercial agriculture	farming animals or crops on a large scale for cash
conservation	protecting the environment
consumable	anything that can be used
consumerism	the belief that it is good for a person or a community to use large amounts of goods and services
contamination	to make something dirty or impure
demonstrate	to show how something is done
destruction	when something is destroyed or ruined
disturb	to interrupt something or someone
diversity	many different kinds of something
domesticated animals	animals that are tamed and looked after by people
dysentery	an infection that causes diarrhoea
ecosystem	a community of plants or animals that are connected by where or how they live
ecology	the relation of living creatures to each other and to their environment
economical	not wasteful, good value

enterprise	a business such as farming or running a shop
environment	the world around us, and the conditions that affect growth and development
exploit	to use something, often to use something selfishly, so that it destroys other things or a way of life
faeces	solid waste from the bowels of animals or humans
fauna	the animals in an area
financial record	a record of money spent and earned
fingerlings	tiny young fish
food groups	the main types of food that supply us with protein, carbohydrates, fats, vitamins, minerals
habitat	the place where an animal or a plant lives naturally
hectare	a unit of measurement equal to 10,000 square metres
harvest	to gather and save a crop from the garden or farm; the crops that is gathered
improvise	to make or do something using whatever is available
innovation	a new way of doing something
integrated farming	to combine different activities to improve the farm
investigate	to find out more about something
irrigation	to supply water to crops through pipes or channels
kangkong	a type of green vegetable used to flavour food
kwashiorkor	a form of malnutrition caused by not enough protein in the diet
landmark	something you can easily see from a distance, sometimes marking a boundary
landscaping	laying out a garden to a set plan
lawn	an area of short grass, near a house
legume	a plant that has seeds in long pods. Peas and beans are legumes
leisure	free time
livelihood	the way in which you earn a living
louvres	slats of glass, wood or metal, in windows
maintenance	to repair or keep things in good condition
malaria	a dangerous disease, spread by mosquitoes, that causes fever and shivering
malnutrition	weakness caused by not having enough food
management	the system of looking after something
manipulate	to handle something skilfully
manure	animal dung used for fertilising the garden
marasmus	a form of malnutrition caused by not enough carbohydrate in the diet
mixed cropping	when more than one type is planted in a garden
mulch	natural material placed around a plant to protect it
nutrients	substances in food, like protein, carbohydrate, fats, vitamins, minerals and water, that you need to keep healthy

organism	any living plant or animal
organic farming	growing crops with the use of natural fertilisers, such as manure
partnership	two people or two groups who work together
perishable	something that is likely to rot quickly, like fruit or vegetables
permaculture	'permanent agriculture', to encourage self-sufficiency and sustainable living by growing a mix of plants which last for a long time, and farming plants and animals together
pesticide	poison used to kill insects and pests
photosynthesis	the way by which green plants use sunlight to turn carbon dioxide and water into food and oxygen
pickling	preserving vegetables in vinegar
poison	something that can kill or harm you
pollution	anything that makes air, water or soil dirty
population	the number of people or plants in an area
preserve	to protect something or keep it in good condition
profit	money left over after expenses or costs have been paid
protein	a substance in food that is necessary for growth and good health
quarantine	keeping people or animals away from others to prevent a disease from spreading
rotation	changing the crops that are planted, each planting season
reforestation	planting new trees in an area where there used to be forest. This helps prevent soil erosion
sanitation	the way in which places are kept clean
self-reliance	able to do or decide things for yourself without relying on others
species	a group of animals or plants that are very similar
statistics	information given in figures
subsistence farming	growing only enough to live on, not to sell
sustainable	to support something and keep it going
symptom	the sign of being healthy or unhealthy
technology	a system where machines or tools are used to make something easier
terracing	a series of flat areas of ground, like steps cut into a hill, used for growing rice for example
traditional calendar	traditional crop calendar used by local people to plant and harvest crops
typhoid	sickness caused by water polluted by human faeces
ventilation	ways to provide a room with fresh air
vegetation	plants in general
viable	successful
virgin forest	a forest that has not been touched or spoiled by cutting down lots of trees
warranty	a promise that something will be repaired or replaced if it is faulty
yield	the total amount of crops or profit produced